FLOWERS FROM BAGHDAD

From the New Jersey Turnpike to Convoying into the heat of Iraq: A Young Lieutenant's Story

GEMMA MCGOWAN

TABLE OF CONTENTS

FLOWERS FROM BAGHDAD

Copyright © 2024 by Gemma McGowan

First Edition

Published by Tactical 16 Publishing

Colorado Springs, Colorado

www.Tactical16.com

ISBN: 978-1-966413-01-1 (paperback)

This book is dedicated to:

All the Soldiers of Alpha Company, 72nd Signal Battalion from Mannheim, Germany, in 2004. "Renegades!!" and "Second to None!"

Foreword

In "Flowers from Baghdad," Gemma McGowan shares her story. It's a story of growing-up, negotiating college, discovering the military, deploying to a combat area, and making bonds with others that can, and will, never be broken. It's a story about a young adult making decisions, and we, the reader, can feel and hear the motivations, emotions, and outcomes of these decisions through Gemma's wonderful writing style. Gemma shares intimate moments when words and small actions generate enduring memories and momentous shifts throughout the course of her journey.

She also shares personal insights into what it is like to be a young female officer starting out in the military. What we learn is that Gemma's quest to be a good leader was not always simple. Yet, Gemma shares how she did it, how she found confidence to create opportunities, take risk, and make decisions for the betterment of her teams and her unit. We also learn that being a junior officer among more senior officers presents challenges. Being a female in a male dominated combat area presents challenges. Being a leader over Soldiers of all ages presents challenges. As Gemma shares her story in "Flowers from Baghdad" these challenges emerge, and we see how Gemma met them and learned from them. This book is a great addition to any reading list for young professionals and junior officers. It is also a book veterans of OIF and OEF will enjoy as Gemma's descriptive and sometimes humorous stories of life in Iraq will likely ring of familiarity.

I have known Gemma since 2007 when she was assigned to 5th Theater Signal Command in Mannheim, Germany. When I met her, she had already joined ROTC and participated in Pershing Rifles, graduated from Rutgers University, moved to Germany, deployed and redeployed from Iraq, and, most importantly, she had found her inspiration and love in leading Soldiers. When I met her, she was still young, but established and self-assured. It was her positive vibe that captured my attention first. Next, it was her continuous involvement in the unit, and at work, to improve the lives of those around her. Finally, it was her infectious enthusiasm and an innate ability to bring people together in ways that are fun which I found most appealing. She had developed a winning formula for inspiring others. It was Leadership 101 and this leadership style had it's roots in the litany of stories shared in "Flowers from Baghdad."

I have been fortunate to continue working with Gemma in numerous ways to support our Soldiers and share our Army story. Gemma replaced me as the President of the General Creighton W. Abrams Chapter of the Association of the United States Army in Wiesbaden, Germany. As I write this, the Abrams Chapter was just honored as the Best Overseas Chapter for its 7th year in a row. Gemma also served as the first female Commander for VFW Post 27, leading the Post to make significant impacts on events and opportunities in the local community. Lastly, Gemma purposefully loaned her grandfather's Eisenhower Jacket, which he wore in WWII, so it can be on display in the USAREUR-AF Headquarters. The jacket carries a message to everyone who visits the HQs; the continuation of selfless service lives on through those who still serve.

At a time when we ask our Soldiers to share their stories so that others can learn about, and maybe even follow, a path into the military, Gemma has done this; she has shared her story in detail. She continues to be the example for others to follow. I am certain that while writing some of the heartfelt stories Gemma shares in "Flowers From Baghdad" tested her confidence and required her greatest poise. Yet, she did it, and she did it because she set out to write a leadership book so that others can benefit from her experiences

When you get to the end of "Flowers From Baghdad," Gemma provides a wonderful summary to her winning formula for leadership by sharing some of her biggest leadership lessons and providing insights on finding her "tribe." Don't jump ahead. Every word in Gemma's story is worth reading.

David Fulton
    President, European Region
    Association of the United State Army

    West Point Class of 1986

    U.S. Army, Retired

INTRODUCTION

Baghdad. 2004. 2300 hours (that's 11 p.m. for all you civilians reading). Even though I tried not to drink before a certain time, sometimes it was just inevitable that I had to utilize the latrine, as they say in the Army, in the middle of the night. The latrines were lined up outside the CHUs (containerized housing units) that we slept and lived in. Unfortunately, war doesn't take timeouts for bathroom breaks. So, everyone had to be in uniform with weapons and ammo in hand. Makes sense though, right? Meeting an Iraqi insurgent outside, in your jammies and unarmed, wouldn't end so well.

Lacing up the boots was always an arduous task - half awake and feeling like my bladder was going to burst. Then I had to grab my cap, because just my luck a sergeant major will also have the same bladder schedule as me and love a chance to square away a young lieutenant. The clock was ticking, and I didn't have the time to get chewed out for lack of hat, so on it went.

I stepped out into the pitch-black night of Baghdad using light discipline. No flashlights were allowed because that would make you a nice, bright target for a mortar or a sniper. Night vision don't fail me now. (I just had to have that apple juice at midnight chow, right?!)

Right in front of the latrine trailer, I stopped in my tracks totally overcome by awe and emotions. Complete silence, complete darkness, and I was all alone - until a massive low-flying Boeing CH-47 Chinook flew right over me. Right over me. If I had taken a stone from the gravel on the ground, I could have hit it. The moment was like in the movies when an alien spaceship flies overhead and you can see every bolt and light. I was staring up like I saw a UFO, too. Beam me up!!

Then it was gone. And like an encore to my private show, next up was the security element, a Sikorsky UH-60 Black Hawk, flying just as low. The door was wide open and there was a badass soldier halfway out of it with a M249 squad automatic weapon. I looked right at him, and he looked back at me as he passed. We made eye contact the whole time. Then there was blackness and silence again. I was alone again. I stood there and was just completely blown away by what just happened. The whole experience not only physically stopped me in my tracks, but I had a moment of self-awareness, clarity, and almost an epiphany. I was overwhelmed with the feeling of HOW IN THE WORLD DID I GET HERE?!?!

So here you go, take these spit-shined black combat boots. Put them on, lace them up nice and tight. Blouse your pants. (Use some snakes or cuff stuffers at the bottom of your pants to look extra sharp). Fall in formation and be a part of my platoon.

This book is a peek into real 'tell it how it was' stories of a young officer straight out of college, to Germany, Kuwait, and Iraq. This is my journey from a suburban town in northern New Jersey, to leading soldiers and finding my flowers from Baghdad: my continuous quest and desire to be the best platoon leader possible, to be a positive leader for my soldiers, and to ensure that everyone came back home safely together.

Welcome to the Alpha Company Renegades!

PART ONE

LIFE BEFORE THE ARMY

CHAPTER 1

DIRTY JERSEY

What can I say? I was lucky enough to grow up in the great state of New Jersey. The beautiful garden state. We have farms, beaches, mountains, and *drum roll* gardens! We don't pump our gas; we pump our fists. Us New Jersites know what a good slice of pizza is, and, yes, we are experts who will give our opinion on this without being asked. Actually, this quote from an unknown hero on the internet really sums it up:

"I am from NJ. I curse... a lot. I say 'yo', and I say it often. I never had school on Rosh Hashanah or Yom Kippur (Jewish holidays). I sure as hell don't pump my own gas. I know what real pizza tastes like, and I know that a bagel is much more than a fuckin' roll with a hole in the middle. I judge people by what exit they get off the parkway. I can navigate a circle--with attitude. All good nights must end at a diner--preferably with cheese fries. It's a sub, not a hoagie or, worse yet, a hero, and I wash it down with soda, not pop. Two words... "mother fucker." I don't go to the beach, I go down the shore. And boardwalk brawls are just a part of the atmosphere. Yes, I drink cawfee. I know that 65mph really means 80. I've always lived within

10 minutes of a mall. When someone cuts me off, they get the horn AND the finger. And they expect it. I am from New Jersey, and damn proud of it."

So, the next time you land in Newark and want to call New Jersey the armpit of America... please remember another motto we have on our border crossing signs: "Welcome to New Jersey! ...Yeah we don't like you either."

In all seriousness, I did love growing up in New Jersey. I grew up in northern New Jersey, toward the east corner of the state where it borders New York state. The town I grew up in, River Vale, New Jersey, had more golf courses than traffic lights. There were a lot of either rich Italians or rich Jews in my area, who mostly lived in big, beautiful houses. If you jumped on a bus in my perfect suburban town, you would be in the center of bustling Times Square in thirty minutes. It is literally a whole new world, so close yet so different.

This is enough about how wonderful New Jersey is, because, believe me, I can go on! The area I came from didn't produce a lot of soldiers, and people weren't desperate to go find themselves in the Army because most of them took over their families' companies, or they had a big college fund waiting for them to go anywhere they wanted to.

I didn't have either of those options, so in high school my father applied for an Army Reserve Officers' Training Corps scholarship for me. I was the founder and president of a few clubs in my high school that focused on leadership and anti-drugs / drinking. I played softball very well and was on the varsity team. With these well-rounded aspects on my résumé and great letters of recommendation, I was awarded an Army ROTC scholarship for college.

My first path took to me to Boston to an all-women's college, Simmons University. I took classes there and did a cross-town arrangement at Northeastern University for the ROTC military science courses.

New Jersey I loved you, but my path was taking me to Beantown.

Chapter 2

ROTC Days

"I am an Army Cadet. Soon I will take an oath and become an Army Officer committed to defending the values which make this nation great. Honor is my touchstone. I understand mission first and people always.

I am the past - the spirit of those warriors who made the final sacrifice.

I am the present - the scholar and apprentice soldier enhancing my skills in the science of warfare and the art of leadership.

But above all, I am the future - the future warrior leader of the United States Army. May God give me the compassion and judgment to lead and the gallantry in battle to Win.

I will do my duty"

- The Cadet Creed

ROTC Life
Massachusetts
I absolutely love Boston. To me it is like a mini-NYC filled with

young college people from all walks of life. I started my freshman year at Simmons College right near Fenway Park.

During the first year in ROTC, they aren't so strict with the freshman, who are called MS 1's (Military Science 1). They don't put a lot of pressure on you and don't want to scare any cadets away. You aren't signed as a contracted cadet yet, so being there is completely voluntary. Even the number of times that you have to show up per week for PT is less. Which was good, because I had a lot of fun with the girls in my own school. We went out around Boston together, and we had a lot of fun and laughs on our dormitory floor. It seemed like I was a typical college freshman having fun with my fellow students.

However, I was different because some days I would get in my Army uniform and go over to Northeastern for Army ROTC classes. Also, some very early mornings I would get in my PT uniform and go there to workout with the cadets. I enjoyed ROTC so much that I joined the color guard team, and we brought in the US flag for their huge commencement ceremony. We were on the jumbotron!

Our program at Northeastern had a training arrangement with the Boston College Army ROTC. Sometimes we would go there for PT early in the morning and get to play sports in the large football stadium. The campus itself was gorgeous! Big stone buildings, beautiful trees, and flowers everywhere. It was a fifteen-minute drive outside the city center, away from what we were used to, and it felt like I took a trip out to the country. You have grass here!? I found this crosstown arrangement to be a very cool experience, and it set up connections with other cadets that I would normally not have had.

The first year was mostly about learning the ranks and the basics of military leadership. I enjoyed it, and I knew that I had a scholarship, so the next few years were going to get serious for me.

I walked over to the ROTC building one of the first days I was in Boston to say hello and get familiar with how to walk there. One of the cadre members there was named Sergeant Captain. He was super friendly and welcoming to me right off the bat. His name was humorous because you have all these college students there trying to learn ranks and then you have a sergeant with the last name of Captain. Some kids thought that was his rank, which obviously doesn't exist. A sergeant is an enlisted service member and a captain is a commissioned officer. I never met anyone in the Army with the last name Captain. But imagine I did, and they were an officer, people would literally have to call them Captain Captain. I don't think I could do that with a straight face. Captain squared ... maybe.

My first day talking to Sergeant Captain, I asked him where the nearest Star Market was, because I was starving with no food. He directed me toward the nearest one, and it was really nice. That's also how I got a nickname from him because every time after that he called me Star Market. Which was cool because that place was awesome. Another cadre member, an actual captain, heard him call me that a few times. The captain assumed that I was the heiress of Star Market, and my father owned the chain.

I only found this out at a very cool dining-in event we had at Faneuil Hall Marketplace, which is rich with history. My father was with me, and he came upstairs to look at the venue. The captain came up to us and asked if he could give a historical tour, which started at the display cases. I wondered why was he being so nice and hospitable to us. Then he mentioned Star Market and asked something about the business. We both looked at him confused, and he said to my father, "Aren't you the owner of Star Market?"

I told him that it was my (oddly given) nickname from Sergeant Captain and my dad is just a lawyer from New Jersey.

The display case tour was over.

At the end of my freshman year, I knew that the school was not the right fit for me. Some of the reasons were the small classes with a feminine twist to all subjects, the classes where girls would come in PJs, and my own personal sprinkle of boy problems at the time. The main reason I had to leave was because the school did not take any of my military science classes for credit because they did not support the ROTC program. So, I did the math with the credits, and I was going to have to double up on classes to graduate on time and take classes that weren't even going to ever show up on my transcript.

Not gonna work. So, I transferred to Rutgers, The State University of New Jersey.

I had to say bye to a lot of friends, and I was really sad to leave good ol' Boston. I made so many close connections with the gals from the women's college, and I was going to miss their friendship dearly. Also, I made some great guy friends at the fraternity house, and we made so many great fun memories together all year. I still spent the entire summer living in the Delta Kappa Epsilon house on the Charles River. They rented out their rooms in the summer to females who wanted to live in Boston. There were a lot of girls from Europe living overseas for the summer and others like me who went to the women's colleges in Boston and wanted to stay the summer while our dorms were closed.

It was a fun summer, filled with roof deck parties with fireworks on the Charles River and nights out at Lansdowne Street going clubbing. I also worked all summer at Fenway Park, serving food at the private parties to all the Red Sox VIP fans. It was a lot of fun, but I knew my future wasn't there. My future was in my new university, and my goal was to graduate to become an officer in the Army.

NEW JERSEY

I stayed in Boston until the last possible day, for Delta Kappa Epsilon's recruiting week, and I pretty much missed every orientation class that Rutgers University had. The fraternity let gals stay there free for a week, and they took us and all the potential freshman brothers out for a week of fun - canoe trips, a comedy show, and another roof deck party. The potential freshman brothers thought, "Wow this is so much fun in this fraternity, and there are girls always around." Then after the week - poof we were all gone.

Since I missed Rutgers University orientation, I didn't know where anything was on campus at all. I had to kind of figure it out somehow and ask people on the street. This, of course, was before smart phones and apps. The original Google search was to ask another person walking by! The first day of class I had at the woman's college versus a Jersey university was such a contrast. I was in classes for the past year with girls literally in their pajamas and fuzzy slippers. Then next to me at the bus stop was a very dressed up Jersey girl with snakeskin pants and high heels, with her hair and make-up all done. What a huge difference.

Toto, I don't think we are in Boston anymore.

Since I was a sophomore now, I was considered a Military Science II (MSII) cadet. It was much more serious, and I was officially a contracted cadet. That meant that they pretty much owned me, I had to graduate on time, and I would commission as a second lieutenant when I graduated. Also, my full tuition was paid for the next three years... so there was that.

The ROTC program at Rutgers University was ranked second in the country at the time, and they welcomed me into their Army family. I truly felt like a Scarlet Knight.

We trained in the local Kilmer Woods, at Fort Dix, and at West Point. PT was Monday, Wednesday, and Friday mornings.

PT was fun and really set the cadets apart from the other students on campus. We were up EARLY for college students. I would be getting up for PT when people were going to bed - or even walking back from a party and then going to bed. Unlike being at West Point, the United States Military Academy, we had to have our own discipline to get our asses out of bed and get to PT. And if a cadet's roommate was not in ROTC, you had two lives to balance and had to creep out of your room early like a reversed burglar.

We would normally do push-ups and sit-ups in the front yard or park, then break into ability groups for a run. Running was always hard for me. I am five feet, two inches and just not built to be a long-distance runner. I would be in the last group, but I did get better over time with a lot of extra running outside of scheduled PT. I was always working hard to stay under weight as well. If you have any curves at all, you will always be over the weight limit and have to undergo the tape test. I hated that test. More on that later.

During my MSII year we were talking to one of our teachers, an older captain, in his office. The whole class was standing in there and also flowing out to the hallway a bit. I was standing in the doorway. He was giving some strange talk about fitness and being in shape.

"Listen cadets, it is really important to stay in good physical shape in the Army. Pass your physical training test, and pass your height and weight tests," he preached while looking pretty borderline passing himself.

Then all of a sudden, he dropped this personal example, "...like for example, if Cadet Gemma loses ten pounds, she could be a Victoria's Secrets model."

I was ... floored. I was embarrassed, and it cut me very deep. I smiled and laughed, but a wave of heat came over my entire body, and I was mortified. Embarrassed that he called me out, I was just standing back minding my own business and listening. Also, I was in disbelief that he used a lingerie company as my life option if I lost weight. Did this just pop into his mind? Or did he already think this before when looking at me all this time? I am thinking the latter to be true.

This comment stuck with me. Still does.

Thank God for the change in the Army because there is no way someone would say this now in the Army environment. You would have a sexual harassment complaint slapped on you so fast it would make your head spin.

Exercise was not the only requirement for cadets, and we had to dedicate other time during the week for training. Every other Friday there was a leadership lab, and we met in the gym to receive in-depth, hands-on lessons on different Army tactical topics. A couple of times each quarter we would have squad tactics training in the woods of Fort Dix. We would also train at the rifle ranges for M16 training and qualification at Fort Dix.

I was so extra 'Hooah' that I joined the ROTC Ranger Club, which did more infantry training than the regular cadet training. We would meet at night in the ROTC building's attic for training based on Field Manual (FM) 7-8, Infantry Rifle Platoon and Squad.

When regular college students were out partying on Friday nights, we did things like tactical river crossings. We did a full silent and tactical ranger river crossing, across the Raritan River in New Brunswick right by the university, in the pitch-black night. I got initiated into the club and was allowed to wear the Ranger Club cord on my uniform. After passing written exams, we also had an induction ceremony upstairs in the attic. We had to stand in a line and the Ranger Club Cadet Commander gave us our cords one by one, then punched us hard in the arm as initiation. I was surprised that this was a part of the club, and I am pretty sure if the actual cadre were there, they wouldn't be allowed to do this.

I joined all these extra ROTC clubs because I wanted to become the best officer I possibly could. I was fully committed and determined to be an effective platoon leader, so I wanted to learn as much as possible.

Our cadet battalion went to West Point to use their woods for our platoon tactics training, and we slept in different places each night on the ground with no tents. I can hardly remember learning much from that long weekend because of some other extreme events that dominate my memory. We had battle buddies, and we stayed with them the whole time. I guess they didn't want to lose one of us since we were in the middle of nowhere.

"Excuse me sir, Cadet Gemma has one question… where are the bathrooms? Port-a-johns?" I asked.

"They are all around you cadet. They are called trees," he replied with a smirk.

Yeah, there weren't any.

Come again? How come no one told us this before we agreed to come on this weekend of fun away from our warm dorms with warm bathrooms? Long story short, my battle buddy and I tried hard to not have to go to the bathroom, well #2, in the woods, but there was no denying it we had to go. We did a long walk of shame ... wayyyy far out from our camp. We would have DIED if any of the male cadets saw us. Die. Dead. Game over. Drop out of school. Witness protection program.

We found a little brook and a rock formation that looked like a small bridge. I went to one side, and she went to the other. We had to shit in the woods like two deer's hiding from hunters and animal predators. I guessed this was normal in the Army. By the way, never ever did I use this great lesson in the real Army.

As IF I wasn't busy enough, I joined the rifle trick drill team. I was the cadet commander for three years. We met every Tuesday and Thursday morning. We spun rifles around our hands and threw riles with bayonets over our heads. Does it get any better than that?!

In the basement of the ROTC building there was an unused drill room filled with so many drill rifles and equipment - buckets and drawers filled with rifle parts, patches, photos, and sabers. There was a rack of dark brown wooden 1903 Springfield rifles with subdued bayonets that we used to practice. Then there was a rack of white wooden 1903 Springfield rifles with shiny silver bayonets. We had the old ascots, hats, and spats as well. It was a vault rich with Rutgers history.

All of this equipment belonged to the Rutgers University Queens Guard drill team that was founded in 1957. They performed their precision rifle drill tricks all around the world. We found photos of the team in 1968 performing at the Edinburgh Military Tattoo in Scotland. Then they kept performing all over the world for the next twenty years.

We found and studied their notes and photos to ultimately copy everything we could. We copied the formations and drill moves from the photographs we found. We also wanted to copy the uniforms they wore. We already had the spats for the shoes, the ascots, and patches in our possession. One thing we didn't have was money.

The ROTC department wasn't funding this at all, so it was up to us to figure out how we could get new uniforms. We were all poor college students, so buying them ourselves was not an option.

The NCO who volunteered to be in charge of us got word somehow that a Rutgers alumni group was looking to make donations to Rutgers' student clubs. I signed up and went to meet them by myself.

Their meeting was held in the Rutgers Club on the top floor. I was nervous to go in, and I had no idea what to expect. I entered the room and saw a group of about fifteen older alumni sitting around a round table. I said my speech about what we needed (uniforms) and what we wanted to do (perform). They asked how much money we needed, and I came up with a high amount of $3,000 thinking they would never approve that much, so they would maybe meet me halfway. After my briefing, I left and hoped for the best.

Shortly thereafter, our NCO told me that they approved the money! The entire $3,000!!! That was, first of all, a lot in 2000 and 2001, and that was a whole hell of a lot for a poor college student to even fathom.

The money went into the ROTC funds, and we bought our uniforms from a parade/flag store. I have no idea what the rest of the money was used for, but I do remember hearing that our battalion commander used it for something.

We looked sharp. We were sharp. And the bayonets were sharp!

We were getting high tech then with graphic arts! For example, we could take images from the computer and get them screen printed on fabrics. What a time to be alive! So, with this new capability, I got drill team t-shirts and windbreakers made for the team. We incorporated the historic Queens Guard crest as well and it looked awesome.

A HARD LESSON THAT WASN'T IN ANY ARMY TEXTBOOKS

When you join the Army as a cadet, they make you believe that anyone in a uniform is respectable and honorable. I actually think they should teach the cadets that this is not true, so they don't find out the hard way in the real Army. The cadre usually only consists of around ten military members, who are (normally) cream of the crop selected from active duty. Being a ROTC instructor is a great assignment. The cadre work at the college, get free classes, and have no chance of a deployment. It is also a good assignment because the cadets worship them and want to emulate their awesomeness.

I had some amazing instructors during my time at Rutgers University – top notch true professional who taught me so much.

However, some cadre were shady. When I was brand new to the program, anyone in the uniform was my idol. There was a man named Sergeant First Class T who offered to drive me to my dorm. I don't remember the situation and why I didn't take the bus, perhaps he was going that way anyway, I don't know. He had a low sports car and the minute we started driving, things got weird. He turned into a total creeper when the car doors were closed.

He was staring at me and at my chest while he was driving and talking to me. He was so damn distracted that he rear-ended a car in front of us on College Avenue! He hit the rear end of the other car so hard that I hit my head and hands on the windshield glass, and the glass cracked. I must not have been wearing my seatbelt. Me, now, would have gotten out of the car and taken the bus after that, but me, back then, didn't.

He dropped me off at my dorm and it was weird. I was so happy to get inside and was creeped out. Then, to my horror, a few hours later he called my dorm room! He got my number off our rosters. Mind you I was barely twenty years old, and he was a crusty old perv. He called over and over and left voice messages on my answering machine. In the final messages he for sure sounded drunk and he was pissed off that I wasn't picking up. I saved his many messages on my old school miniature cassette tape and brought it in to the colonel.

The colonel was REALLY mad! He automatically got on the phone and started calling people. In about a week or so the perv was gone for good, and I never saw him again.

After he was gone, the female admin sergeant took me in a back room to talk. Normally she was a really hard tough NCO, but now she was crying to me and was saying sorry. She said she was sorry that she didn't stop him, and he was doing the same to her. He would always say how there is so much ass walking around (ewh) and he couldn't believe it. I didn't know what to say to her because nothing happened to me. I got the feeling that she thought he sexually assaulted me already... and that he did assault her.

My hard lesson was that people are people regardless of what they are wearing, and you cannot trust everyone. The Army uniform does not make you a super-hero; you most definitely can be a shitbag. It was a hard truth, but a much-needed lesson.

CADET TRAINING

Advanced Camp

After the third year, also known as the junior year in college, cadets have to go to a camp where they are tested and evaluated in every way possible for six weeks. Doesn't that sound like fun?!

Cadets were literally tested for everything (health, physical fitness, leadership, stress, etc.) and then ranked among their peers, and also by their peers. The environment was similar to basic training that new enlisted servicemembers go to, but we were all college students coming with three years of military science and leadership knowledge. We were not being "broken" by drill sergeants because we already had military bearing and we all were there in order to be commissioned as lieutenants in another year. ROTC alone required a lot of dedication, and we willingly spent our college summer break in shitty situations, lacking sleep, in the woods, in WWII barracks, and under leadership pressure. All the while our friends from college were in Cancun drinking cocktails in coconuts with tiny umbrellas.

Cadets were assigned to platoons and those were basically the only people we had interactions with during the entire six weeks. Even though there were other groups of platoons around, we never mingled with them. We slept in old, open-bay barracks (which means it is one large room with rows of bunk beds on the side) with two floors. There was one bathroom on the bottom floor that we shared in shifts. (More on that nightmare later.) The top floor was filled by all males; I honestly don't think I ever saw what it looked like but I assumed it was a mirror image of our floor. Someone told me they used to fart in front of the big industrial fan, so maybe that is also why I didn't go up there.

On the bottom floor, the first few bunks in the back were all the females. We had about ten females in the five bunks. The rest of the bunks were the overflow of males. I was on the border of the males and females, and there were no curtains or anything like that. We didn't hang out in the bays, so it was okay – we just went there to sleep. Overall, the people at camp were my first interactions (more than just meeting) with anyone outside of the greater Northeast. My bunk neighbor was from Kentucky. Maybe I shouldn't use the word neighbor so warmly because I'm pretty sure he hated me. He told me that my tattoo (which is the size of a quarter) was sin and I was going to hell. He was completely serious, and he believed it.

"Your body is your temple and if you are tattooed the bible says you are going to hell," he matter-of-fact informed me without me asking.

"I didn't know my lucky shamrock was going to lead me to a life of eternal damnation, what a bummer..." I thought to myself as I stared at him from my upper bunk.

This was my first interaction with someone from Kentucky and off the bat, it wasn't going so well. No one ever told me I was literally going to hell before. I mean, I am from Jersey, and we tell people to 'go to hell' all the time...I mean at least three times in rush hour traffic going into the city... but this... this was new to me.

If I was at regular basic training, I would have had all of my items thrown on the ground and my sheets ripped off. However here at Advanced Camp, I didn't make my bed one time the entire time, which is very rare for Army training. I kept the sheets folded in squares that that they came delivered in and under the green wool blanket, which I spread out on top and tucked the corners under the mattress. I slept on top of the wool blanket and used my poncho liner as a blanket. Then in the morning I would just need to get up and put the liner in my locker and not mess around with tucking in sheets on a top bunk. Like I said, never made my bed once.

Such a rebel.

There were some days we had free time, which was mostly on Sundays, but we had to stay right by our bunks and not wander more than a few steps away. I would sit on my bunk to write letters home and write in my journal. We had no music or electronic devices; just pen and paper. On some days the other girls from the south would sit on their bunks, hold hands, sway back and forth, and sing about Jesus. This was also very new for me and no one I knew did that.

Now, two of these same girls who were swaying back and forth and singing about Jesus acted completely different in the bathroom. The first girl, who was Hispanic with shoulder length thick light brown hair, would walk around topless in the bathroom. She would brush her teeth... topless, and load laundry... topless. She did this because the door was right there and it would open when people went in and out and the guys out there could see her. Seriously.

Once I said, "Um.. you might want to move when the door opens."

She replied very carefree, "Oh it doesn't matter." Then I could see her looking into the mirror to see when the door opened to see if someone saw her. This boggled my mind.

I found out ... in the showers ... that the second gal, who was a small framed, short black female with black slicked back hair, had a tattoo. I hated the showers, and I don't understand the point of the "training" for it. We had three shower heads and ten girls, and we had to shower at the same time, with about ten minutes to shower... because we were in a big rush to get somewhere?!? And, by the way, never in my Army career did I need to know how to share a shower nozzle with anyone, let alone a shower room with ten girls. So, one of the Jesus singers had this massive tattoo across the pecks part of her chest, practically from armpit to armpit, I swear, that read 'SEX MACHINE' in big block letters. I wouldn't have believed it if I didn't see it myself with my own eyes.

The first few days were all medical testing. Everyone went as a platoon to the medical facility, and we got everything tested that was possible to test or inspect. At the end of the day, we got brown bag lunches with chips and a candy bar. We were allowed to eat lunch slowly, sitting on the grass, and it was one of the few times where we weren't rushing for something for no reason. Or as we call it in the Army... hurry up and wait.

After we got back from the last medical day, the commander got a call, and a few cadets had to get on the bus and go back to the medical facility ... because they had chlamydia! When they got back, they had to go in the commander's office and call their significant other and tell them the 'news.' Let's look at the positive side here... hey, free call home!

Imagine being the commander having to stay in the office and hear the conversations. I imagine it went something like this, "Oh hey babe, yeah Advanced Camp is good. Yup got to use the phone already, pretty great right? Yeah, I miss you too. So um, actually the reason I am calling is, um, yeah I have chlamydia and you might, too. Okay well, I am out of time, gotta go bye bye now!"

At Advanced Camp, either you were a cadet or you were cadre. The cadre were the bosses, and we were bottom of the barrel nothings who just needed to do what they said. Even though a lot of the people working there literally just got commissioned a month before this. Meaning they were cadets a month and a day ago.

There was a lieutenant working at the camp for his first assignment before he would go on to a regular active duty to a real unit. I think the first time I saw him was when I was on staff duty in the middle of the night. Staff duty was a shift that was always covered by someone. It was shitty and unnecessary to me. We were already tired out of our minds, and we had to go to an office and wait to see if a phone rang. As if we knew what to do if someone did call. We would just go wake up an important cadre member. So basically, we were cell phones.

So, this lieutenant worked in the office most of the time; he was tall, skinny, and had blonde hair. I was in no way flirting with him, and flirting/attraction never crossed my mind for a split second. I am a friendly person, and, unfortunately, sometimes it is taken the wrong way. More so when I was younger, because now I am a hardened old person. That shift at staff duty sparked some imaginary shit in his head because after that he would look at me or come up and talk to me. Like I said, as cadets we had no voice and no rights.

He would come up to me and talk to me casually when I was standing in line for chow. This was really strange and not done. I don't remember the things he said, but I for sure picked up the vibe that he was interested in me somehow. And I couldn't turn it off because we weren't allowed to talk back really.

One day my platoon was practicing marching on a basketball court, because God forbid, we do nothing for a few minutes. The lieutenant came up to me and whispered in my ear (this I cannot forget) and said, "I always imagine what you are wearing under your uniform, and I picture it being sexy and made of lace."

Ewh. (And remember the Victoria's Secret comment from my teacher?! ... It brought back that terrible feeling again.)

In my head I thought what the fuck!? But I couldn't talk back to him because he was... cadre. I didn't say anything to him. I just ignored it and wanted to pretend that it just didn't happen. Later on thinking about it, I was actually shocked that he thought women wore lingerie under their uniforms, especially at camp. We wear sports bras and Hanes Her Way underwear. It got me thinking, do other guys think about what women are wearing, too? Gross man. I assure you most girls at a camp are definitely not thinking about what is under a guy's swamp-assed uniform.

So, I thought about this for a little while, and a little voice inside me felt like I needed to tell someone. In the beginning of camp, the very nice lieutenant colonel in charge of the camp told us he has an open-door policy, meaning we didn't need permission to come to him and talk about anything. That somehow stuck with me and I made the decision that I was going to take him up on it.

I went to the same building where we had staff duty. Thank God home-slice wasn't there at the time. I sat in the lieutenant colonel's office and looked around at all the plaques and trophies on the wall. It was nice to sit in an office and not in the barracks or be on some obstacle course. I felt like I was visiting civilization. He was very nice to me from the moment I came in and asked to talk to him. I told him what the lieutenant said to me about the underwear, etc. You could practically see him boiling with anger and he started to look freaking furious; but he assured me that he wasn't mad at me.

"I am not mad at you, cadet, I am glad that you came to me and told me," he said.

"Also, I promise that it will never happen again, and you should feel safe here at Advanced Camp," he assured me with a very serious facial expression.

He was right... I should feel safe there. I didn't need that extra bullshit on top of all the stress they were throwing at us daily. Well, the cat was out of the bag, I went back to the fart-filled barracks, and I didn't know what was going to happen next. What's done is done now. Karen talked to the manager.

The next day I saw him walk past me with a big pouty face and looking at the ground while he walked really fast. He was carrying some flags, and they were slipping, and he was juggling them – but like an angry juggler who just got yelled at. I don't remember seeing him again; they probably gave him an administrative job somewhere far away from cadets. I was glad I didn't have to avoid him anymore, and honestly, I didn't care if he was mad about it. I hope it stopped him acting like that to other subordinates for the rest of his Army career.

SOME OF THE active duty cadre working at the camp got some serious power issues while being there. They tried to make it a basic training environment; you know the one with real drill sergeants? But it wasn't the same and even then, being so young, I could see the awkwardness in some of them.

There was one old school master sergeant who was totally wacked out and hardcore. He was the old Army! He must has seen some shit in Vietnam because this guy was not right. He walked around always being like the drill sergeant from 'Full Metal Jacket.' He was 24-7 intense and pissed off.

We were doing PT one very very early morning outside on the grass. I was so tired and not feeling it. I swear to you, the one time in my entire time there at camp I wanted to half-ass it for a few seconds was when we were doing push-ups. When you do a push-up, you look to the ground and you have to go down low enough to break the plane, meaning your chest dips lower that your elbows. Everyone was looking down and we were knocking out like forty to fifty of them on our own. Everyone was looking down including whoever was leading PT. I was dead tired, so I did these little, half-assed push-ups and didn't break the plane because I thought no one was watching. Nope. Guess who magically appeared from around the back of the building? Yup, the crazy master sergeant. He immediately ran over to me and started yelling at me. He got down next to me on the ground to see if I was breaking the damn plane.

"Break the plane, cadet! Why are you not breaking the plane?! Why are you being lazy?!" he yelled at me, with his head right next to mine on the grass.

I was shocked at a few things... 1. This was the first time I ever half assed on purpose and look what I got, and 2. ... where the hell did he come from?!? He appeared out of nowhere. It was as if he was watching us from the bushes or from behind the back of the building. I then did some extra push-ups for my little lazy moment.

Luckily, he didn't remember me after that or constantly make me pay for my previous lazy moment. I was not labeled as a 'shitbag cadet' in his permanent memory.

Me being me, I volunteered for the class color guard since I had a lot of experience from being the drill team cadet commander at Rutgers. I went to the first try-outs/practice and, low and behold, guess who is in charge of the Color Guard?? Yup Full Metal Master Sergeant.

I knew my drill, so I was confident. I was trying out to be a rifle bearer. If you are on the right side, it was easier because you hold the rifle in the arm that you normally would hold it, the right arm. However, the color guard was mirrored with a rifle bearer on the left side as well, and all the flags in between them both. So, the rifle bearer on the left has to hold the rifle on the left arm and salute the rifle with that arm; I was used to saluting with my right hand, and it was a little awkward for sure.

I could do both because I ate and breathed drill team back at college. I was on the right side, then to my left were two tall flag bearers and then on the end the rifle bearer holding it on his left arm. The master sergeant was watching us intensely as we marched. So, then we did 'present arms' and the other rifle bearer fumbled it a little. Eek! He got a warning from master sergeant. On the next time around for 'present arms,' he fumbled it again. With his usual intenseness, the master sergeant went over to the rifle bearer and, no joke, punched him hard in the upper arm!!

"Do it again!" he says.

Holy shit!! This color guard just got really serious. I was thinking, "Please don't mess up rifle dude... please don't mess up rifle dude."

Try outs were over. I didn't make the team because they told me I was too short. I am too short in regular life, but also the six foot, four inch flag bearers made me look even smaller. I was really disappointed when they told me that. It isn't my fault I am short. I told my sergeant major who was there also from my college; he said it was bullshit, too.

There was also a female officer there and when she spoke in front of the entire group, I picked up a real shitty vibe. She was loving being in charge and she was playing some hard-ass role but not even well. When we were out in the field we only had the port-a-johns for bathrooms, and she was yelling at everyone because someone put in a maxi pad in there. So, she is yelling at the top of her lungs and says, "The next person who puts a maxi-pad in the port-a-john, we will find you, and all the females here are going to scoop through the port-a-johns, find it and take it out."

I thought it myself, "Is she even hearing herself talk right now? Is this what leadership is to her?" She became some psychotic power monster. We were all about twenty-one years old, and she was talking to us like that and trying scare us. I thought how easy it could be for her to just say something like, "Hey ladies, the port-a-john emptying sucker thing gets clogged, so don't put any maxi pads in it." But no. She threatened us to all be elbows deep in port-a-john shit and piss. She was acting like a huge bully in order to get the cadets to listen to her, when good leadership could have gone a long way.

The first few weeks we resided in the barracks area mostly, and we took day trips to different ranges and obstacle courses nearby. We all did these tasks already in our ROTC training at school, so now we were being watched and tested on it all. We did M16 ranges, grenade ranges, multiple obstacle courses, group problem solving obstacle courses, bayonet training, and a lot of weapons cleaning.

I did fine in the different courses because Rutgers ROTC really set me up for success and prepared its cadets very well. I only failed one day and had to repeat it and ... I cried in my one-man single shelter tent. (Yeah I know, there's no crying in the Army.) I failed the night land navigation course. This was done in the very thick woods of Fort Lewis, Washington. I knew my land nav skills, so I was shocked when they called my name as one of the people who had to come back to try again. After they called my name, I took my points to the land nav cadre and asked them to show me where I failed.

To save time in land nav, we were taught at Rutgers to find a spot on the map, like a road intersection, that is close to your point, instead of counting your steps from the start to the first point. So that is what I did. Daytime Land Nav was easy breezy. While other cadets started counting from the start, I took off on a slow jog to the point, just like how I was taught. Unfortunately for night land nav, the spot where I thought it was my first road intersection was actually a tank trail. How in the hell was I supposed to know what a tank trail looked like? Well, I will tell you what it looks like... it looks like a road, not a trail! From that location on all my points were thrown off whack. Hence the failure. I was already exhausted and now frustrated with the whole tank trail bullshit, I crawled into my little tent and cried. Not my proudest moment! I cried over the frustration and also that I had to COME BACK and do it all over again.

The next night the bus came to the barracks for all the failures to take us way out deep in the woods again. Instead of hundreds of cadets stomping around the trails for the same points, there were about twenty of us. You have no idea how thick and dark these woods were. Now I had to go out all by myself and hope to God that I didn't meet a hunter or civilian out there (which they warned us about!) and become a famous rape and murder story. However, on the bus, one of my platoon mates, one of the cadets I could actually tolerate and was nice, came up to me.

"Hey, do you want to go get our points together," he asked.

"Oh, hell yes!" I replied excitedly.

Maybe he was scared of being murdered or running into bigfoot, too? Regardless, I was super grateful, and we went to his points and my points together... and we passed.

During the last week we slept out in the field and did squad tactics with blanks and laser gear – MILES, the multiple integrated laser engagement system - to show if we were shot or not. We took turns being in charge, and when we were in charge there was a cadre literally following us around during every move and writing on a clipboard. If we weren't in charge at the moment, we had to execute whatever the cadet in charge wanted done. Even if it was a stupid plan and chances were we would get "killed," we still had to do it.

We all got briefed on potential dangers we could encounter in the woods. The first one was straight out of all of my nightmares: black widow spiders. Let me state that again. Black. Widow. Spiders. Thank God there were no smart phones back then, so I wasn't able to Google image them to see photos and what the bites do to you. Both terrifying.

The second thing they told us to avoid was small camps in the woods. They said we might see a small circle with a fire in the middle and some garbage around it. They said these were made by mostly homeless people. Well naïve me thought that people were just coming in the woods to camp out in a random spot and make a fire. Years later after watching all of 'Breaking Bad' and a documentary about the meth problems in Washington state, I realized these were actually... METH LABS! Why they didn't tell us the truth, I will never know.

Also, the woods at Fort Lewis had HUGE ant hills that were taller than I am. They had ant highways that would come out of them. You would see these dark lines from far away and then get closer to realize that they were ant highways with ants going in and out of the huge mound. While doing squad tactics in the woods, you have to be careful getting down in the prone position to not have your leg going across one of these ant highways because your entire leg would be covered in black ants. Some of them will bite and it really hurts!

When newly enlisted soldiers go to basic training, those soldiers can bond through shared adversity. At the end of basic training, the platoon feels like a tight knit family who have grown so much with each other. They keep in touch for years and call each other "battle" (as in battle buddy). The training for soon-to-be-officers was not this way because of all the evaluation competition among each other and class ranking. We weren't friends; we were competitors. Yeah, everyone heard me cry myself to sleep in the woods, and everyone showered together but still, no one kept in touch after that. I didn't even exchange an email or address with anyone, and I am the friendliest person I know.

I was happy as hell to get out of there. The last day we were there they let us walk further on down the dirt road where we had just lived for the past month and go to a sorry excuse for a PX. It was in an old barracks building that had basic snacks for sale. Even with the weak shopping selection, we felt grown and accomplished to be there! They kept us at the barracks until the last minute before our flights home so that cadets couldn't get drunk at the airport and miss the flights. Because apparently this happened all the time in classes before us!

For me, training wasn't over yet. I wasn't going home. After a cadet goes to the Advanced Training, they have the opportunity to go to a real Army unit for two to three weeks and shadow a real lieutenant. Coming into the training, I didn't have a slot to one, but an NCO from my Rutgers ROTC was at my camp as well. He was kind of looking out for me the whole time, which felt really great. He found me one night and told my leadership that he needed to talk to me. I was thinking... oh my God, what did I do?!

He brought me over to the smoking area and said he had news for me. I have never been to the smoking area before, so I looked around. "So, this is where they go," I thought to myself. I also got the same 'I don't belong here, but try to blend in and look cool' feeling like I did in high school when I would go to the secret smoker's area to accompany my smoker boyfriend.

"Cadet, I got you a slot to go to Cadet Leadership Training after camp," he declared.

The slots are very hard to come by and very sought after.

"Wow! That's amazing, thank you!" I said while giving an awkwardly cool thumbs-up like Fonzie, trying to blend in with the smokers.

"Cadet, you have one of two choices. One, go to Korea to an air defense artillery unit. Two, go to Fort Gordon, Georgia to a signal corps unit," he said while then looking for me to reply.

Oh, I need an answer there on the spot. "Umm.... Umm...." I said as I pondered.

And then I made a decision that for the rest of my adult life I will wonder if I made a stupid mistake. Yeah, I probably did. I picked the signal corps in Augusta, Georgia. Why, you ask? (I ask myself the same still) Because I was really determined and focused on getting into the signal corps, so I thought that this would help me learn and also help build my résumé to get selected. Looking back, I probably would have been selected for the signal corps without that extra training, and I gave up a pretty awesome opportunity to have an all-expense paid trip to see Korea. I didn't have that spark to travel or see new cultures quiet yet. I was a poor college student just trying to survive and get commissioned! That's all I really thought about. I was trying to be the best officer that I could become.

I left directly from the Advanced Training in Fort Lewis, Washington, to Fort Gordon, Georgia. I called my parents from a payphone at the airport and said, "I am going away for another two weeks!" No emails and no cell phones back then.

When I arrived at Fort Gordon, I was brought to the hotel on base where the lieutenants stay when they go to officer training. I didn't know it yet, but I would stay there also when I was a second lieutenant the following year. It was SO nice to be in the hotel room!! My own bathroom, the bed, the air conditioner, and most of all I was by myself! I was exhausted from traveling, and my bags were lost. I had no extra clothes and no sneakers. I just wanted to pass out from being tired. My lieutenant that I would be shadowing called me and told me she was picking me up early in the morning for physical training. Oh, sorry chick, I don't have any sneakers, no can do.

Luckily my bags were still in Washington. The next morning, she still picked me up and I went to PT in the battle dress uniform (BDU). This is when the shit show started.

The lieutenant I was assigned to was totally crazy. She was white, tall, and also built bigger than the average woman, with long sandy blonde hair she wore up all the time. The next two weeks I would embark on an insane rollercoaster ride with her and her social life that I didn't sign up for.

I got to formation, and I had to stand next to her and just watch everything she did. When I did end up doing PT, I sweated like crazy after running. I could not believe what humidity did to your sweat glands. The first day I had no sneakers, so I didn't do any PT. I was introduced to the platoon, and they told everyone to treat me like a platoon leader and salute me. So, I come from this camp where I am a piece of nothing who isn't allowed to speak to anyone or talk back, to being saluted by forty-plus year-old men. All righty then.

My lieutenant was from the Midwest, and she had a funny accent to me. I had a funny New Jersey accent to her. Also, when I told her about my sneakers in person, she busted out laughing about it.

"What is a sneaker!? Are you going to sneak around with it?" she asked me while cracking up.

Apparently where she came from, they call them gym shoes. Okay, but what if you aren't going to a gym? I like sneakers better and I shall never refer to them as anything else.

I came right in the middle of when her love life was starting to bloom. She had just met a guy ...introduced by her mom, (okay, that's okay), over a group phone chat, (like an internet chat room, but on the phone), and she had never met him in person. He lived about four hours away from Fort Gordon. She told me she was sorry that she couldn't take me out that weekend because she was going to meet him for the first time. They were already dating, and they said they loved each other already. She was nervous about going because she was still a virgin. Here I am twenty-one years old, she was about twenty-seven, and I could tell that this weekend trip was a bad, bad idea. But she wasn't asking my opinion, she was just letting me know I was on my own so I better order take-out pizza.

The next Monday rolled around, and she called me and told me she wouldn't be coming to PT that morning, but she would still pick me up and drop me off. (Damn my sneakers and I can't get any PT breaks around here.) She looked and sounded terrible, but didn't say much when she dropped me off. It was the officers' PT day so I was there with all the officers to include the battalion commander, a lieutenant colonel. I joined them on their run and did well. I even called cadence, and everyone was surprised that I wasn't shy to do it. I couldn't believe how my running had improved and that might have been the best running day of my life. Good timing.

I went back to my room to change and get ready for the day, and the lieutenant picked me up and spilled the beans about what was wrong with her. She went to the medical facility... because there was a piece of her hymen (she thought) and it was, you know, there. Also, she was sore, you know, there. I didn't know how to take this information, nor how to react to it. I wish she actually lied to me and said she had a migraine or something. Also, I was thinking, could you imagine if all the officers knew that was the reason why she missed PT this morning? I also didn't know why I was involved with this. All I wanted to do was hang out in the air conditioning in my room and watch Jerry Springer.

The rest of the week was great. Everyone showed me their jobs in the company. I learned that my last few days there would be spent in the field with them. I wasn't ecstatic about that (remember my Jerry plans) but at least I would learn more about the signal corps.

It was nearly Friday and the lieutenant asked me for help. She asked if I would go shopping with her.

"Shopping, yeah sure. I love to shop. What do you want to get?" I asked.

"For... I don't know... things," she said followed by a giggle.

I was so sure that she was insinuating about sexy outfits she wanted. I didn't ask for details, and away to the store we went with me riding shotgun. To my surprise we pulled up to a Walmart. What is happening?

She went to the birth control aisle and bought every type of birth control possible. Things I didn't know existed. She was going to use all of them at once. She was shopping for her next trip that weekend. Oh, no, she was driving all that way again. I asked why he didn't come to her, and she said he didn't have a car or couldn't leave his state or both. Right... are we at the Les Misérables barrier because that is a huge RED FLAG.

Cadet training was good for me, and I was really glad I was able to go there. In the field I was able to see the signal equipment in action. Not having a clue how it really all worked, I was fascinated that it did. I learned how all the assemblages were able to talk with each other, and how hard the cable dawg soldiers worked to bind it all together. I saw how the satellite was the backbone and pulled all the data and phone service into our area in the woods. It was great for me to see this firsthand, rather than being shown a PowerPoint slideshow or in a textbook.

I ALSO LEARNED some life lessons, like how my lieutenant was completely distracted by her phone chat relationship. Or how a soldier, whose wife just had a baby, got my phone number from the roster somehow and called my hotel room late at night asking me what I was doing and if I wanted to meet him. I assume his wife was in the hospital still. Before the call, I thought he was a really squared away soldier, and I learned again the hard way that just because you wear a uniform, it doesn't mean you aren't a piece of shit.

I met an NCO who was living in the barracks because his wife was trying to kill him. Yes, literally trying to slowly kill him. I wouldn't have believed the story either unless I met him in person myself. He had a thick southern accent, was bald, and pretty round in shape. He was one of those guys with a big round belly and skinny legs but could run forever. Like their round belly is a keg that supplies all the running energy. Anyway, he would come back from PT every morning and drink a big glass of Kool-Aid from a pitcher in the fridge. His wife picked up on this routine and took the opportunity to... put RAT POISON in the Kool-Aid. He was getting really sick, and tested for some type of poison, so they came to his house and tested his environment. Lessons learned here: change up your routine and don't marry someone insane.

I met with the battalion commander there and she told me that if I didn't get branched signal corps to call her and she would get me in. Wow, doesn't that sound great and like I made some really good connections for my, hopefully, future in the signal corps? Must have been worth tossing that trip to Korea, right?

Being a contracted cadet changes the kind of person you are. I was now really in the Army. I was for sure going to be commissioned (unless I really screwed things up). An example of how it changed me was clear when I went to the Port Authority in Manhattan after camp - a place I have been so many times in my life. The same route from the same bus station. I came down the big high escalator to the main station, like always, and saw the American flag and the New Jersey flag in flag holders... but they were in wrong order. The American flag was on the right side. As a cadet in the color guard, I could not leave it alone like this. I went up to the flags, swapped them, and then saluted the flag the right way. A Port Authority cop was watching me; I looked back at him and told him, "They were wrong..." and I walked away.

9-11

Back at Rutgers ROTC, I was a big bad college senior/MS IV in ROTC, and I had just spent time in the 'real Army.' It was a beautiful clear sunny day, and I was sitting in my small senior seminar class for my major in American studies, and the teacher came running into our class and said, "A plane hit the World Trade Center!"

Everyone was in shock. I didn't really understand the severity of it yet. There was no smart phone to pull up photos or any televisions in our area. Honestly, I don't know how they found out so fast. It was on TV, and someone must have called the office and told them. (I had a flip cell phone but only used it on the free nights and weekends.) A girl next to me started crying uncontrollably because her parents worked in one of the towers. I was still in a weird state of shock or denial. I didn't know what to do or think. Class was canceled and I took one of the many RU buses back to the main campus. There was a weird vibe in the air. What was weirder was the silence. There were no planes flying, no city buses, no trains. Silence.

I went directly to the Army ROTC building, and many people were out on the front porch. We all sat there just being confused. One of our teachers, a captain, was on the phone and practically ghost white. He was trying to call his wife who worked in one of the towers, but he couldn't get hold of her. It turns out that she slept in that day, was late for work, and was stuck in the subway underground for hours, but totally fine.

Then our battalion commander, a lieutenant colonel, came out to the porch and said, "A plane just hit the Pentagon!"

Then breaking the silence in the air, a fighter jet flew right over our heads headed to New York City.

What the hell was happening?! And at this point we still had no TVs or images.

I walked back to my room (at the frat house) because I guess all classes were canceled the rest of the day. I walked into the common area and that is where I saw the first images, which was the live news on the TV. Oh my god!

The news was also playing the plane crashing into the tower, over and over again, and videos of the collapsing tower as well as images of the person they thought was responsible: Osama Bin Laden.

I couldn't understand why someone in Afghanistan would want to attack and kill us, all the way over here in America. Also, what did this mean for my future going into the Army?

In the past, I would often go to New York City at night with my college friends. For weeks later the lower part of Manhattan was closed. We didn't know that though. We found out when my friend and I drove to the Holland Tunnel a week or so after 9-11 and pulled up to a very confused toll lady. She said, "We are... closed."

Oh.

Honestly how would I find out these things if I didn't watch the news much?

It wasn't for a while later that I was able to visit Ground Zero and see all of the rebuilding You could still see the impact of the blast left on the surrounding buildings. We went to a deli that was open right in the area, which somehow wasn't completely damaged. Inside and outside there were many memorials for the firefighters and first responders. I felt solemn and it didn't feel right to eat my sandwich while standing there.

As cadets we would have found out what Army branch we were going to be selected for right before fall break. However, the general who died in the Pentagon was in charge of all the assignments, and that whole office area was also destroyed. So, we didn't find out what branches we were assigned until the spring and almost the end of the school year. The West Point cadets get first pick. Then all of the Army ROTC cadets are ranked in order based on how well they did at camp, college GPAs, and rank in their classes. It is called an order of merit list. The assignments staff go down the list and cadet's "wish lists" to try to give everyone their number one choice, until the slots are all gone, or based on the needs of the Army.

We all stood around the conference room and our battalion commander came around one by one to shake our hands and tell us what branch we received. It was a big moment for us! It was the first information about our futures in the Army - where we would go for our officer training, and also a good indication of what bases we might be stationed at afterward. Besides the needs for the Army that year, it was a good indication of how well you did overall, if you got your first choice or not.

When our battalion commander came up to me, she shook my hand, and I lost the ability to hear any other sounds for some reason. In slow motion she said to me... air defense artillery.

Wait, what?

What??

Everyone was jolly and celebrating. I was in a state of shock. What just happened? How did I get my third choice?! I had tried my absolute best and it just wasn't enough. There was no way I wanted to go into air defense artillery. I added it as number three because all women had to add a combat support branch on our top three. I picked signal corps, military intelligence, and air defense artillery, in that order. I felt like a failure, and that I was fooling myself that I was good enough. My future looked very dim and there was no way I wanted to do anything with firing artillery in the sky. While I totally respected the branch and all my mentors who were ADA, I knew it just was not for me.

Everyone knew I was all about going into the signal corps. I mean, I was one of the few who was picked to get a slot for the cadet training, and I just spent weeks with the real Army Signal Corps. My battalion commander knew I was devastated. We went downstairs to the admin NCO and asked if we could submit for another branch. There could be some leftover spots, or someone else wanted to swap. I did the paperwork which put me on standby for the transportation corps, and I was close to tears.

Then I remembered about the battalion commander in Georgia who told me to call her if I didn't get signal corps! My battalion commander thought it was a great idea! We went into her office to make the call. I called the main number to the staff duty for them to connect me. And who answers the phone: the same NCO that called me in my hotel room late at night! What are the chances? Out of hundreds of soldiers from the battalion he had to be the one on staff duty. Damn, I didn't need this drama now. Whose karma was this coming back? Mine or his? Anyway, I acted like everything was hunky dory, especially since I had an audience on speaker phone. He connected me to the battalion commander, I was practically holding my breath. This was my one chance to undo this ADA nightmare I just entered.

"Hello ma'am, remember me? Remember when you told me to call you if I didn't get Signal Corps? Well, here we are!"

Welp, to my disappointment, she started telling me how great ADA is and how there are only a few women in ADA and how I can really stand out. Here was another life lesson learned. She was just trying to be nice when she said that to me. She didn't have the power to change my branch. Why did she say it to me then? To see my gratitude and reaction to her generous offer? I will never know.

Fast forward to a year later in Fort Gordon, I saw her at an officers luncheon. She saw me and avoided me.

At this point in my life, this was one of the first times I felt completely devastated. I felt like I failed. Everything I was doing for years, the best I possibly could, was not good enough. I was in such a daze leaving the building I forgot that I drove there and that my car was parked along the side street. (That was a nice ticket and fine later.) I walked back to my room, missed my classes that day, and just went to bed. When I get depressed or overwhelmed, my body shuts down, I get tired and just want to sleep.

The next morning, I get phone call on my awesome Nokia cell phone.

"Oh my god, it's the battalion commander," I thought. "Why is she calling me?? She never calls me!"

She said, "Cadet, there has been a big mix up... and actually you are the only good news to come out of this. There was a mistake with Cadet Command, and they sent out the wrong list. You didn't branch ADA; you were branched Signal Corps."

"Umm... what? Am I still dreaming"? Literally felt like I was still dreaming like in the cartoons. I had to rub my eyes... pinch myself... smack my cheeks.

The conversation was quick because she had to call a list of cadets. Some of my friends went out the night before with their families and had a big celebration because one was selected for the medical branch that he wanted so much. Another group of friends were happy to get infantry, also their long-time wish.

The rumor going around was that on the master Excel sheet, lines were deleted and all the columns got shifted before that file was emailed out.

The crisis was diverted, and I felt a huge weight lifted. So, I did get Signal Corps, and I would be going back to Fort Gordon for officers training. However, the first duty station was not given to us yet.

So, one day I was chilling out in my room in the fraternity house I was living in then. There was a handmade bar left over from one of the past brothers, and I made it into a very cool computer desk. I used one of the barstools as my desk chair, and I hung my Mardi Gras beads off the back of the chair. I was enjoying the new fast LAN connection, chatting on AIM chat, downloading music on Napster, listening to it on my Winamp with cool skins, and checking my AOL mail.

I got an email ("You have mail!") that almost knocked me off my barstool draped with Mardi Gras beads, and if I really knew how much it would change my life, I should have really fell over for dramatic effect:

"Dear Cadet, you are assigned to Mannheim, Germany, after Officer Basic Course."

Auf Wiedersehn Dirty Jerz

But not so fast! First I had to go to and complete four months of Officer Basic Course at Fort Gordon, in Augusta, Georgia. I had been there already, and I knew how different it was in the south, so I was prepared.

Graduation from Rutgers University was bittersweet. I was excited for my future in the Army, but I also was going to miss this cadet/college life. College was fun because I bounced around two different worlds: college girl life and then Army girl life. My last days of college, I was running around the fraternity house with my best guy friend at the time, with ruck sacks on our backs, and camo on our faces. We made MRE bombs and threw them out the windows of the frat house to scare the girls in the neighboring sorority house. We laid back with our rucks on our back, drinking out of our canteens, and watched war movies in the common area of the frat house. I knew life wasn't going to always be like this and I was going to miss it. It was time for the 'real world' and the 'real Army.'

However, inside I knew I was ready for the Army.

We were commissioned as second lieutenants before the actual university graduation ceremony. I commissioned with a small group. I had my favorite sergeant major, who helped with the drill team, conduct the traditional first salute. After my commissioning, my father had a party with my entire family and longtime friends at the Rutgers Club. We had Revolutionary War reenactors lined up greeting everyone at the door, and before the dinner started, they came in to say congratulations and a big 'HUZZAH!'

When I walked across the stage at the Rutgers University commencement ceremony, they announced my name as "Second Lieutenant Gemma… " that made me feel very proud and excited for what was ahead in the Army. I was ready to be a platoon leader with soldiers.

ROTC Swearing in ceremony with my Grandparents and Mom

Ice cream truck ready for our convoy

Chaplain saying goodbye to me.

Renegade driver, ready to depart.

Military Convoy route signs.

Selfie with my driver on the way to Baghdad.

Victory Base Motor Pool, Baghdad.

Palace on Victory Base, Baghdad.

Renegade TACSAT site from the view of the radio tower.

Me with 'Victory Over America Palace' in the background.

'Victory Over America Palace' entrance.

Saddam's Flintstone Village, view looking out one of the houses window.

Group Platoon photo at my farewell gathering.

Alpha Company (acting) Commander Headquarters. Official
Company 'Hot or Not' contest on the wall in the back.

One of Alpha Company's final formations in Kuwait before going
home to Germany.

Me and my WWII Vet Grandfather on a VFW float during a Memorial
Day Parade in our hometown.

My Grandfather's jacket and other items, on display at the
USAREUR-AF Headquarters Building in Wiesbaden, Germany.

AUSA Chapter Members in General McAuliffe Square during the
NUTs weekend in Bastogne, Belgium.

PART TWO

IN THE ARMY NOW

CHAPTER 3

2LT

There I was at my commissioning ceremony repeating the words, "I Gemma, having been appointed an officer in the Army of the United States, as indicated above in the grade of Second Lieutenant, so solemnly swear that I will support and defend the Constitution of the United States against all enemies, foreign and domestic, that I will bear true faith and allegiance to the same; that I take this obligation freely, without any mental reservation or purpose of evasion; and duties of the office on which I am about to enter, so help me God". – Army Officer Oath of Office.

A great feeling of pride and excitement came over me after saying the Officer Oath. It was an incredible feeling thinking about my future. I am getting so close to my goal of being a great leader of soldiers.

SIGNAL CORPS OBC: GET YOUR FREQ ON.

My Signal Corps Officer Basic Course was in Fort Gordon, Georgia. This is where Signal Corps second lieutenants learn the basics in all different types of communications, because there is no telling what kind of actual unit you will be assigned to. When officers get to their first units, that is really where they learn the more up-to-date technology and the platoon's capabilities.

Besides one trip in college to New Orleans, going to Fort Gordon was my first experience in the south. Everything seemed so slow paced. Like literally as if everything was being played in slow motion. And there I was, a very fast-talking Jersey girl. When I talked to people you could see the steam coming off their heads in overload. People would say to me all of the time, "What did you say?? Talk slower! I can't understand you!"

I was a stranger in a strange land. I also noticed that I was the only person wearing black all of the time. Everyone else wore bright colors like orange, yellow, and red. Which are all colors that I didn't even have in my wardrobe. People from New York City tend to wear black and considerate it really chic. Apparently in the south they only wear black at funerals.

I had just spent my senior year having fun in NYC and now I was pretty much just ready to get through the officer training classes and not fail. While everyone would go to Atlanta on the weekends, I had no desire to join them. I hung around the base, and I hardly even went into Augusta. I am sure I missed out on some nice experiences, but I was just not interested at the time. I was always keeping my head above water with the classes and passing all of the technical tests. It was ALL new to me. It could have been in Chinese, and it would have been the same learning curve.

Even though I stayed to myself there were numerous guys trying to get with me. During my cadet time at Fort Gordon, I met a male lieutenant who was friends with the lieutenant I had previously shadowed. At the time he seemed pretty cool and chill. When I got to Fort Gordon as a lieutenant for OBC, we bumped into each other again. We went out to dinner, and I brought a fellow female classmate with me. The guy told me many personal and shocking stories about him and other women recently, and all these jerk things he pulled on these women. I was shocked but I realized he was talking to me like one of 'the bros' and I was completely okay with being in that role.

A few days after our dinner and his nasty story, he showed up at my room after going golfing in the Augusta heat and humidity. Not like the outcome would have been different, but he was sweaty and smelled like serious funk. He came in and was talking to me while I was at my computer (probably on AIM and listening to Winamp) and he tried to kiss me. I said, "Um... no thanks." He was like, "Come on why not?" I guess I wasn't in the bro-zone like I thought I was with his terrible dating stories, and he also was so stinky, the whole situation was nuts. Luckily, he did take no for an answer, and he left... hopefully to go home and shower.

At OBC I felt like a fish out of water in all senses. In the classroom, in the state, among the students. At lunch my refuge would be going back to my air-conditioned room, making mac and cheese, and watching Jerry Springer or Maury. I probably could have made better use of my time, but that was my escape.

I woke up every day at some ungodly hour, like 0445, to be at the field at 0515 for PT to start at 0530. We would run in the dark. (This is where I unfortunately learned when zero dark thirty was) We would do pushups and sits ups in the dirt... but mostly we ran in the dark. I hated running in the dark. On weekends I would also go running on this big sandy road around the base. Like Baywatch but shitty. I sweated so much again in the Georgia sun in a very different way than I would sweat in New Jersey. I then learned why we ran in the dark.

I still wasn't the best runner, but I tried my hardest! I am five foot, two inches and have short legs, and I was running along with tall men, whose legs were nearly my whole height. People who weren't in the military might not understand that running in formation is a totally different beast than running on your own to music in the gym or in the park. You have people with all different running abilities and heights together running in step. Yes, someone is calling cadence and that's fun and all, but when you feel like you just can't keep up, it still sucks. If it is a big formation, your section will have the accordion effect, so it will be slow, then get really fast and you almost have to sprint. You must stay in your spot, so you must keep up with the pace. The pace setter is the person who is up front and to the right. Normally this will be the average runner of the group, or someone who is very good at getting the right pace/time. I am short so I was always somewhere near the front.

The worst thing that can happen is you 'fall out' of the run. This means you can't keep up with the pace and you are falling behind. So, everyone keeps running and you slowly filter back until the whole formation runs past you as you look like a loser who can't keep up. Then you keep running and try to follow the formation in the back. There are always road guard runners to stay behind you, and if there are a lot of people falling out, then someone else will leave the formation to help out the road guards and police up the stragglers. It is a terrible feeling to fall out of a run.

Some of the guys in my class were mean to me because I wasn't the fastest runner. I fell out in OBC once and one guy said to me as he passed, "Yeah, you will make your privates proud." What a mean thing to say. He was such an asshole. That made me so upset I really couldn't run fast after this because I had the overwhelming feeling of wanting to cry.

I wasn't as vocal as I am now, so I chose the long passive-aggressive payback, instead of just confronting him and telling him off. My friend and I went to Spencer's Gifts at the mall and bought some ridiculous bumper stickers. Late at night we went to the parking lot and put them on his car. There were some rainbow pride stickers (which was not allowed in the Army back then) and some 'honk if you're horny' type stickers. If we had cell phones back then, I would have loved to take a photo of that. We wondered how long he was driving around before he realized they were there.

The fact I still remember him saying that really shows how much it hurt me. I didn't know what kind of platoon leader I was going to be. I was never in the Army before. This ass clown was prior service, so he was in the Army before as enlisted. I was in contact with my college non-committal boyfriend, who was a big bad first lieutenant at Fort Bragg already, and he made me feel better when I told him about it. He said, "Well he is right... you will make your privates proud."

And eventually I did. So, you can suck it Thompson.

Like I mentioned, I wasn't very vocal and not so verbally sassy yet, but I did a few things like a night vigilante... or like one crazy Jersey Bitch. We had an exchange student in our class and the first female officer from Mongolia. She was pretty bad ass and cool, but she didn't talk a lot.

OBC was a wild learning experience for sure. We had officers from Chad and Congo, and they hated each other. There was one bigger guy, I think the one from Chad, who would go to the military clothing store to buy all Special Forces, Ranger tabs, Airborne wings, unit patches, and put them anywhere he wanted on his Chad uniform. Everyone was too scared to say anything to him, so they let him wear them. The other guy from Congo was more chill and easier to talk to. He told everyone that his girlfriend in Congo had HIV, but it was okay because he wore two condoms when they... you know. We were flabbergasted by this tidbit of information. (Also remember in 2002 HIV was looked at very different than it is now, and now it is more manageable with medication.)

Back to the Mongolian lieutenant. Each foreign lieutenant had a sponsor who drove them to PT, class, and other activities. Basically, the sponsor helped them whenever they needed it. Lord tell me why someone thought it was a good idea to match her up with a male; I will never understand that. We had so many females in our class. So, she didn't speak much, but one day she opened up to a small group of women lieutenants that her sponsor cornered her inside his pick-up truck and started kissing and touching her. I believe she was married as well, and this would have been devastating if anyone from her country or family found out. She made us swear not to say anything.

What I am about to tell, I do not condone or suggest anyone try this at home. I was really mad at that guy, and I was a twenty-two-year-old hot head from New Jersey, who thought a good solution would be to smash out the taillights of his pickup truck.

That will teach him.

So, I came up with a great idea: Put my baseball bat (of course I had one) inside my dress uniform bag, zip it up, carry it on a hanger, go to the parking lot, and smash out his lights when no one was looking. And that was what I did. I kept a piece of glass as a souvenir for the lieutenant. She must have been like, "These women are crazy as hell over here," but after that she gave me two cute Mongolian dolls in a nice paper box. If I had Google translate I would have told her, "Don't worry, if he tries that shit again I will smash his headlights next."

We stayed busy at OBC. All of the lieutenants had to do a certain number of community service hours and also take a class leadership role. I volunteered with a group a few times at a veterans retirement home for their bingo nights. Sounds nice right?

Well, it wasn't nice, it was actually horrible.

All the people were fighting with each other, calling each other names, and calling out people that were cheating.

"He has bingo cards hidden under his lap blanket!!" one elderly veteran yelled while pointing his finger at fellow veteran man across the table.

I didn't even know you could cheat at bingo.

Then they brought out imitation Jello cups.

They were yelling at the nurses, "These are gross!! Where are the ice cream cups!?"

I was scared.

For the (mandatory) leadership role, I gladly volunteered to help with planning the Signal Ball that we have at the end of class. I was pretty much in my element party planning and enjoyed that part of the school. Other than that, I was suffering along with everyone else by death by PowerPoint, then doing a brain dump after each test.

My partner, Lieutenant M, was an extremely motivated and boastful person. He was like the poster person for the class, and he made it known he was proud of himself. He had his own business cards with a long list of all his degrees and certifications. We went to meet businesses and other people for potential sponsorships and he would pull out his business card, making me look like a dud next to him. In the end, we had a great ball, people had fun, and the intent was met. Everyone got a nice Army coin as a table gift. I still have it on my desk at work to this day.

Years later, I found out online that Lieutenant M killed himself. I was shocked then, and I still am. He seemed so motivated and happy with himself when I knew him. I didn't keep in contact with him for long after school. We were MySpace friends, and he had his rank in his display name, which at the time I thought was odd but very like him. I don't know the details of his suicide, but it really made me sad to think about what happened to him, when I always felt that he was headed for greatness. And now he is gone.

The months at OBC were eventful to say the least, and somehow, with the grace of God, I passed all the signal flow and technical tests. In the gift shop at Fort Gordon, they had a mug that read, "Happiness is Fort Gordon in my rear-view window." Not to sound too harsh or to pick on Fort Gordon, but I most definitely had that feeling when I left the gate.

I was ready for to join my Army unit in Germany, and I was eager to be a platoon leader. I wanted to make an impact, and this was not the place to do so.

MANNHEIM

I was twenty-three years old when I landed on my own at Frankfurt Rhein Main airport in Germany. There was a German man cleaning up and sweeping, and in my head, I thought... "Oh, that's a real German." That sounds weird, but this was a totally new experience for me. I never met or saw a real German before! It was more of a reality check that I was really really in Germany!

Mannheim was the place to be! We had so many soldiers stationed there, and actually all over Germany at that time. This was before the big soldier drawbacks in Europe and base closures. It was cool to be an American in Europe. We could wear our uniforms off base and didn't have to worry about terrorism then. We all drove around with our state flag banner hanging down from the rear-view mirror. (You better believe I hung a New Jersey one with pride!) The dollar was almost equal to the euro, and we were living large. The Germans loved coming on our base to shop and go to our clubs, and we loved shopping in town and going to their clubs. Most all of my American guy friends had a German girlfriend... or two.

I was a single officer, so I was allowed to live off base. At the housing office, they gave us a list of houses to look at and we could choose where we wanted to live. So, I had just been living in a hotel for the past few months in Georgia, and before that in an almost condemned fraternity house that looked like the Animal House movie, so my standards were not very high at that point. The housing office recommended an apartment for me in an adorable wine town at the foot of the mountains in an area they called the Bergstrasse (the Mountain Road). At the first look I took it! Little did I know how much that decision would shape my life so much later on.

I loved my apartment. I never had my own place before besides dorms. I kept it in perfect condition, and I collected all my European treasures/souvenirs from my many trips and had them on display.

There were a few bus companies in the area that did awesome trips all over Europe for Americans. There were information stands set up all over the base in public areas with all flyers of the trips and the dates. Some were overnights in hotels but most of them were express trips. You leave on Friday night, arrive at your location Saturday morning, spend all day and evening there, then leave Saturday night so you come home again late Saturday or early

Sunday. No hotels were involved, and you slept on the bus. There were bus trips to Paris, London, Berlin, Munich, and Amsterdam, just to name a few. The price was under $100 for the trip, and you couldn't beat it. Bring your pillow, snacks, MP3 player, and you are set!

And if you were unlike me and you had a decent car and not a hooptie – also known as a beater car - you could go on great road trips all over Europe. Americans stationed in Germany got to fill up at American gas prices, even on the road. We had our own gas stations on base with US gas prices, but we were even able to buy gas coupons there and use them at any Esso gas station in Germany. So, when you planned your road trip, say to another country, you mapped out the last Esso station by the border and you filled up your tank. The savings was significant. It cost about double, if not more, to buy German gas.

The gas station on the Army base would be the happening place to be right before a four-day weekend. Everyone would be filling up their gas tanks and then also buying some gas coupons. There was one woman who ran our gas station in Mannheim. She confused me, since I was new to the Army, and I will tell you why. She wore an Army jacket, with Household 6 as the name, and she wore her NCO husbands rank on the uniform. I was wondering... is that allowed? What is actually going on here? She would walk around like she was a military person. Household 6 is a call sign on the radio, and people used that as a joke when talking about their wife at home... as if they are one that really runs the household, and you have to check in with them for permission. I used the word wife because I literally never heard a female soldier call their husband Household 6.

I really wondered if some new privates would go there, see this woman decked out her Army uniform, and question if they should be standing at ease... like you would for a real NCO with that uniform on.

We worked hard on the weekdays, and we played hard on the weekends. The younger soldiers who didn't have so much money to spend on bus trips or any other kind of travelling would have barracks parties. This was off limits for any officers. You don't hang out in the barracks… especially a female officer. The senior NCOs and the first sergeant would be the ones in charge of making sure no extreme crazy things happened in the barracks. The squad leader and platoon sergeant would have to make several barracks visits and checks. They did this to mitigate any serious things happening that would be elevated to the officer leadership.

The young officers would hang out together. We would mostly go out to Irish pubs, restaurants, or travel together. Then the officers and NCOs who had families were also traveling a lot and going to local fests.

"Work like a captain, play like a pirate."

I was assigned as a signal platoon leader to the greatest platoon in the Army: the Alpha Company Renegades, Second Platoon, "Second to None"! We were a newly configured platoon which was called a heavy data package. We could deploy anywhere and provide communications as a long-term solution for the units in the area. The other platoon was the light data package, which could provide the same services on a slightly smaller scale, but they would jump site along with a unit that was on the move to different locations.

Our platoon had an eighty-five-foot satellite dish connected to the satellite van, connected to a single shelter switch (SSS) - which was the center control of the network - and that was connected to a data center. Each squad in my platoon was responsible for the different parts of the heavy data package.

GARRISON LIFE

In the Army world you are either in garrison, which is your home base where you are stationed, or you are deployed, which is on an assignment for about a year to either a combat zone or a non-combat zone. Some gray areas are if you are 'in the field' which is kind of like garrison, but you are out in the training area and you don't sleep at home... you sleep in the field. The field is not really a field, you sleep in tents and (in the Signal Corps world) you set up your communication equipment and trucks to test if they work or not. Then the unit will be qualified as ready to deploy to support a combat unit if needed. We work out any kinks there may be before we deploy for real world situations. Basically, going to the field sucks and no one really likes it, but everyone knows it has to be done. So, when you are in the field you focus on what has to get done with hopes that they will end the exercise early and you can go home and sleep in your own bed. Meaning, it's all work and no play in the field. Our mentality was, "get this shit done and over with."

So being in garrison is the most relaxed place a unit can be at the moment. This is also where a lot of the problems happen! The first sergeant, also known as Top because they are the top of the enlisted soldiers in the company, is the one in charge of ALL soldier business. He or she knows everything about each soldier at any given minute. They know their where abouts, marital status, PT score, run time, weight, allergies, where they live, what they do on the weekends... you get the idea.

As an officer you don't know the details of each soldier, but you support the first sergeant when they need you to do something. Smart officers will just take their word for it because you don't question Top!

The first sergeant and the company commander are a team; likewise, the platoon sergeants and the platoon leaders are a team. The company leadership gives guidance to the platoon leadership. But everyone knows that Top runs everything.

Accountability is the most important thing for an Army unit. It is always the first thing you do each morning, throughout the day, and at the end of the day. Then on the weekends, Top is still responsible for any soldiers acting crazy or missing from the barracks.

You don't have formations in the field much because people are on shifts and spread out all over the place. You don't have formations in a deployed setting because that automatically shows the enemy who the leaders are, and you are just asking for a sniper shot to the head of your commander. Or being all clustered together in a formation is a very easy target for a mortar to fall right in the middle of everyone.

However, each morning in garrison there is a morning formation where Top receives a report from all the platoon sergeants about how many are present for duty and those who aren't... where they should be. Lord help you if you are called out as 'failure to report.'

As an officer, only the company commander was tracking my whereabouts really, and if I wasn't there the soldiers would think I was off on important officer business somewhere.

One place you could be was sick call. Sick call was the early morning appointments in the medical clinic for soldiers waiting to see a doctor. Back in my time, the soldier would have to come to formation first, then go to sick call after being accounted for. Some people used sick call as an excuse for not doing PT that day, but if you were actually sick this would be your only option to be excused from PT. There wasn't an option of going to lay down during PT because you didn't feel well; you literally need to see a doctor and get a sick slip for even a cold or a tummy ache. Some soldiers were so hardcore they would not be caught going to sick call ever.

Sick call was an interesting experience for me a few times, because I look so young, like a young private, people thought I was an enlisted soldier at sick call. Let me tell you, I got to feel it first-hand how young female soldiers got treated like absolute shit.

In Fort Gordon at Officers Basic Course, I went to sick call one time because I had killer skin hive allergies for the first time in my life. (I think I am allergic to Georgia.) We had to go to sick call in our PTs, which was just a sweatsuit with no rank. Since I had a baby face and no make-up on, I got treated like I was about twelve years old until they pulled my medical record.

There was a sergeant in charge of the sick call waiting room. He was the big dawg in the room because all of the other soldiers in there were just out of basic to attend Advanced Individual Training

(AIT). I stood against the wall, realizing the shitshow I just signed up for and how I didn't want to be there. But since I was a student, too, there was no time during class to make an appointment at the medical facility to get allergy medicine, so it had to be done in the morning.

The sergeant told me really sternly, "Sit down, right there, in that chair!"

"No thanks, I'm good here," I said.

Now being stern turned to anger and he said, "Excuse me, what did you just say?!"

I stared back at him unfazed, holding up my part of the wall.

Before his head exploded, he realized I was a second lieutenant and just left me alone.

I thought, "Damn, this is how you get treated in AIT if you are enlisted?"

Sick call in garrison was not so bad, because people sort of knew each other and also, we didn't have any military schools on our base. The doctor would see you, prescribe you some medicine... most likely Motrin... then if you were lucky, you would get a slip to take back to Top that put you on "Quarters" status for the next three days. Or if it was even more serious, they would put you on a temporary profile, which means you can't do PT or go to the field. The profile would be detailed about what you could and couldn't do. No leadership would ever mess with a profile from a doctor, and you don't break your profile or do what it says you can't do. You can add that to the list of the responsibilities of Top: knowing who is on profile and for what.

The medical facility was a busy place on base with a constant flow of people coming and going. A lot of military members get as much dental and medical care as possible while they are still in. Sometimes when they join the military, it is one of the few times in their life (or first) that they ever saw a dentist or doctor. Every military member has to go get medical and dental checkups or they will come up red on the MEDPROS report. And if you do come up as red, your first sergeant will send you right away to get checked so the report will be green again. This reflects the unit's total readiness to be able to deploy and fight.

I remember when I was getting ready to ship out to active duty, I had my mandatory and first dental appointment. I was super nervous because I knew I had a molar that was not right and more hollow then it should be. The dentist came in and was kind of young and good looking, which didn't make me feel less nervous or embarrassed about it. Then he tipped me back in the chair and one, two, three, he fixed the tooth of shame. He handed me my dental records, and I was on my merry way. As a new military member, this made me feel very taken care of.

I didn't go to the doctor much growing up unless I was very sick or in pain. Now in the Army, I felt the freedom to go for anything that I felt like and also knew that I would get free medication or supplies as well. This is something I never took for granted. Dry eyes? Go to the troop medical clinic and get drops. Allergies got you down? Go see a doc and get meds. One time I went there with back pain and for some unknown reason, they gave me a shot of Demerol in my buttocks. So then after that I was walking around the clinic drunk. I was slurring my words while talking to people that I met in the hallways. I don't remember how I got home. I guess I waited there for it to wear off. Who thought that was a good idea?

Each garrison has a few key services that is similar on all Army bases. Of course, the most popular is the PX, or post exchange. It is owned by Dallas-based Army & Air Force Exchange Services (AAFES) which has 2,700 facilities all over the world. They even are in combat zones! It is the only shopping that is possible when you are out in the middle of nowhere. They also have tactical trucks with a mobile PX in the back.

The PX is always a good time. It is like a small Target, where you can get pretty much everything there. The selection isn't huge, but they have what you need and also military items as well. Then the outside of the PX there is a food court with typically the same vendors at each place. Of course, Air Force bases will have more options and better everything... but the Army can at least go there as well and pretend it is ours for a few hours.

The commissary is owned by Defense Commissary Agency (DeCA) and they provide discounted groceries to the military. Being stationed overseas, the commissary literally gives you a taste of home - at the same price you can get the same food and household items that you are used to in the States, in another country... and that's huge.

Each post will have some sort of dance club. Back when I was in the Army, posts in the States had places for the young soldiers to go party, an NCO club, and an officers' club. So, everyone could go 'let their hair down' and not do it in front of people they are in charge of or ones that are in charge of them. It also prevented any fraternization, where people in the same chain of command, and also enlisted and officers, could not party together. This prevented any problems in the ranks where others will think you give special privileges to some people because you are bros in real life.

In Germany we had two clubs on the post that we could go to, and they were for everyone. Overseas there isn't as much land to expand or build on the post, so they don't have as many clubs as in the States. The two clubs in Mannheim were a lot of fun, and a lot of German gals met their American Army boyfriends at these clubs. By the time I was in the Army, the base had a gated entry after 9-11 so any non-military person had to get signed in. However, if you knew one person, they could sign in a small group of German gals. I didn't go very often to the clubs; I only went for a promotion party or a unit farewell event. ... but I heard all of the stories!

There was a club called the Pirates Cove on Taylor Barracks, but it was shuttered right before I got there. Someone told me that a girl was murdered in the club, so they closed it down. It was some love triangle or jealous feud. They also tore the building down after some time, and it was in the parking lot of our dining facility.

There was The Cove on Sullivan Barracks, which was an old German officers club. It was open at night-time as a club, but I never went there during these times. I only went there when we had several hail and farewells and unit events there. The building architecture was old and very unique, made of dark wood, and gothic looking in a way. You kind of expected a sophisticated vampire to be living in the attic of the place. When I was in charge of one of the unit events, I had to go there to meet with the manager for planning. He told me a little bit about the history of the German officers club. He told me that they found a bunch of old German handguns buried in the yard right outside the club.

So of course, I had to ask, "Do you ever see any ghosts?"

He told me, "So at night I have to wait until everyone leaves after the club closes. First thing I do is make sure the building is empty and then I lock the doors. One night after locking up I heard a door slam and someone running away. I thought it was someone drunk that got left inside, but when I followed the running footsteps, it only led me to a locked door."

Spooky.

Another reminder of a dark past was embedded in the structure. If you looked hard enough around the buildings that the units were in, which used to belong to the German Army, you might find some swastikas in the architecture. For instance, there was a building on Taylor Barracks in Mannheim which belonged to the military police where we had to register our privately-owned vehicles. There was a big wooden staircase, and at the bottom was a very nice thick wooden hand carved banister. At the end of the banister, the wooden swirls came together and the final design was a big swastika. I guess Americans tried to paint over it a few times, and they thought the layers of paint would fill it up, but it was clear as day still there. This was shocking to me to see. I thought about who must have gone up these stairs in the past... and here I am using the same stairs to register my nineties Mustang. I have never seen one before in person, and while growing up the topic and that symbol seemed so taboo. I have wondered what happened to the banisters after the post closed.

Did it just get smashed into the rubble? Maybe this wasn't the history they wanted to keep anyway?

For the record: *Assistant S1 isn't a real job.*

My first job when I arrived in the unit was as the assistant S1 (admin officer), which wasn't a real position. I was in this position for a few months until Alpha Company came back from their exercise-turned-real-world mission in Israel, then I would take over a platoon as a platoon leader.

I didn't know what I was doing really. I would follow direction from the captain who was the actual S1. I watched and learned what the team of soldiers did in the S1, like process promotions and leave requests. The S1 was also in charge of all of the evaluations, and they had to brief the battalion's admin status at the command and staff meetings. Not really super exciting stuff, but it was good to learn. This was also my first work email I ever had, and I used Outlook for the first time. The Global Access List (GAL) was like cutting edge technology, where you could actually search for a name in the GAL and send someone an email. With all my years of using AOL, I was pretty much a pro at learning Outlook. Who I would write to, I had no idea. Maybe to some people in the motor pool so I didn't have to walk down there. Maybe we sent slides and Excel sheets to each other. Soldiers didn't have their own emails; only people in leadership positions did. We didn't write personal emails to each other; we would save that for our AOL, AIM, or Yahoo chat when we got home.

The S1 had a senior NCO who was in charge of the section. He was very old in my eyes, and he was about to retire. He was for sure from the 'old Army.' He was quiet and didn't talk so much to anyone. It was apparent that technology was not his friend at all for basic tasks. One day he burst in my office and desperately asked me, "How do you recall a message?"

I replied, showing him, "So, go to the sent message, click recall message."

Then he left.

Why was he so frantic about it, though? I have never seen him talk so much to me in my whole time being there. So, I go to my inbox and there is an email from him, then above it a request to recall the message. Do I click on the recall request email? Hell, no I don't. My life in S1 is boring as shit, there is no way I am going to miss whatever he wanted to pull back so bad.

So, by mistake he forwarded a wrong email to the entire S1 section. Did I scroll all the way down to read the whole conversation? Hell, yes I did.

A married senior NCO was having a conversation back and forth with some young soldier somewhere, and it wasn't his wife. He was talking about how the weather was bad, and we got an email that all non-essential people could go home. That was the email he thought he was sending to everyone. Instead, they started talking about the cold weather, and how he needed to keep his bald head warm. Then the conversation turned to talking about keeping his 'other head warm' and how she can do that, etc. EWH! Recall failed here, and now I want to forward it out of my mental image forever.

He got a written counseling from the S1 captain for the email. Basically, a slap on the head...oh I mean wrist.

Around the same time, I had another 'Outlook gone wild' instance, but this time with a lieutenant colonel. To put that in perspective, I was a BRAND NEW second lieutenant, and the only lieutenant colonel I would come in contact with is my battalion commander. I was on a Neckar River boat cruise that was hosted by First Command, an investment company that caters to the military. It was an awesome time! No-cost boat cruise, dinner, and all you can drink wine. I went with a few other lieutenant friends, but we scattered around the boat to mingle. I found a table with a nice couple, he was a lieutenant colonel and his wife, who much younger than him, was from South America and didn't speak a lot of English.

Not a problem, because my high school Spanish was on par in those days. Especially after all the wine I drank, I was habloing like a pro.

I wasn't used to chatting with a lieutenant colonel like this, but he seemed super cool. Look at me, hanging with the big dawgs! Lord knows what we talked about, just basic things like where we grew up, how we liked Germany so far, nothing crazy, and we didn't take any photos together. We didn't exchange information, and we ended the night with, "Good luck and have a nice life."

A week or so later, to my surprise, I get an email at work from the lieutenant colonel! I guess he remembered my name and looked me up in the high-tech GAL. He said that he was thinking about me the other day when he was passing our base in Mannheim, so he wanted to drop me an email. Sounds nice so far...a little odd for a lieutenant colonel to write to a second lieutenant like this, but we were river boat buddies now. Then the zinger at the end of the email: he writes a P.S. but in Spanish... "I have some photos that you might like to see. Let me know if you want me to send them to you."

What the actual hell?!

There might have been a shitty Babelfish translation website, but I grabbed the nearest Hispanic specialist in the orderly room and confirmed if my high school Spanish teacher taught me well or not. Yup. It says what I think it says. I was shocked. I sure as hell didn't want to see any of his photos. (Did I almost receive the first ever Outlook d*ck pic?!) Maybe they were innocent and photos of German castles, but I wasn't about to find out. I was intimidated by him because of his rank. I was scared he would get mad at me and maybe retaliate somehow, and I was worried I would bump into him again. Luckily, I didn't.

This new Outlook life was wild.

BECOMING A PLATOON LEADER

I left the S1 shop and it was now my time to take over as a real platoon leader. This was the position that I have been training to be in for about five years as a cadet and also as a young 2LT. I was in my S1 office, and my future platoon sergeant, SFC Penton, came in and introduced himself to me. I was so impressed how professional and squared away he was. When he came in to meet me, he made me really excited about becoming a platoon leader and working as a team with him.

I left S1 and reported to Alpha company the following next days, officially as the platoon leader. I watched The Soldiers looked up to him so much. He didn't ask for people to do unnecessary tasks, so when he did the soldiers executed. He taught them to be the best at any task you did, not matter how big or small, and to take pride in it. He often told them, if you were going to take out the trash, you be the best trash taker-outer that anyone has ever seen. We were soldiers. We were professionals.

He had a special way of motivating and inspiring soldiers that I have never seen since. For example, he gave out the "highly coveted 2nd Platoon Jalapeno cheese spread award", from MREs to a soldier who stood out and achieved excellence. He taught the platoon to always respect me as the platoon leader, and I really appreciated that. He would speak to them bluntly, but in a joking way tell me, "ear muffs Ma'am", and I would put my hands over my ears, because it was something a little rough for the soldiers and not pertaining to me. Everything thought this was funny, and it was also a display of respect to me as the officer of the platoon.

We were a tight platoon who enjoyed working and training together and it was all because of his leadership and guidance. He truly inspired every single soldier in the platoon for the rest of their lives. Even in the soldiers spare time, they would still want to build comradery with fellow Renegades. We also bonded with fun trips around Europe!

SPAIN

The bus trips were just the best during these times! Living up to my 'Julie the Cruise Director' nickname that my platoon sergeant gave me, I planned a bus trip to Spain for the whole company. The bus was 95% our unit, and the others were from the community. The tour left on a Thursday night, arrived on Friday, and departed on Monday morning for an evening return. It was awesome because almost everyone in the company went on the trip, and the ones who were married brought their spouses. It was a double-decker bus, and everyone piled in with their pillows, beer, and snacks. The cost for the trip was extremely reasonable because it included the drive, hotel, and a tour of Barcelona: around $120. I mean... come on.

The tour company took us to a beach town in Spain about an hour outside of Barcelona. The driving time from Mannheim to near Lloret de Mar was twelve hours without stopping... so on a bus it was about fourteen-plus hours. We didn't stay right in Lloret de Mar; we were at the less exciting, less party central area of the beach, and that was completely okay with me.

Even though we didn't sleep so well on the bus, we were in Spain!! We were all so excited. It was everyone's first time there. We had a lot of Spanish speaking soldiers, so they were able to communicate for us, and also it was a great time for everyone to use their high school Spanish.

"¿Donde está la biblioteca?"

I was so ready if anyone was looking for a library.

We arrived, checked in our rooms, and then everyone dispersed. It was a super cute beach town with so many little shops, cafes, and bars along the water. It is common in Spain for women to go topless at the beach, so this new information was buzzing around to each person they saw from our unit in town.

The next day there was a tour of Barcelona included for everyone. It was about an hour drive from our beach town. The same bus that took us there from Germany now drove us into the city. They gave a quick windshield tour of Barcelona, then dropped us off at a spot and told us to be back there at a certain time. Barcelona was awesome! It didn't matter that it felt like it was 150 degrees in the sun... we were in freakin Barcelona! ¡Qué divertido!

We got back to our town right before dinner, and we had time to go out to the clubs as well. We were all sun burned from the open-top, hop-on hop-off bus tour, but that didn't matter because were motivated to go out and soak up as much of Spain as possible.

The dance clubs were fun, and they played a lot of dance techno music that we weren't used to. The drinks were cheap and, in my opinion, then, the Sangria looked much nicer than it actually tasted. I was expecting some sweet, delicious, fruit explosion, but it was kinda bitter and the whole time drinking it, I only looked forward to getting some chunks of fruit at the bottom of the cup. Well, looking back we probably got the cheapest version possible from the places, so it wasn't a fair assessment.

Sunday fun day, beach plans and exploring in the town again, then going to clubs at night. It was actually the last night we had in Spain because in the morning we were going to board the bus to go back to Germany. The longgggg bus ride home. Well before that torture came, everyone was for sure up for the last opportunity of nightlife.

I did go out for a while, but I was tired from the whole weekend and my sunburn was on fire, so I went back at a normal time. When I was going up the stairs an American guy who was staying there was going down the stairs. He says to me something like, "My room is upstairs on the next floor just so you know." I said, "Oh ... okay?" And then went to my room. It didn't click with me for like an hour or more later that he was inviting me to him room. How naïve was I?! No one ever said anything like that to me... at least that I noticed. *double locks hotel room door*

Rise and shine came too early, and it was time to get on the bus. I packed up my bag and headed to the meet-up spot. Some people were hung over, some people didn't sleep much if any at all, and some people were coming right from the club. In theory it was a good idea... stay up all night and sleep on the bus.

If the Army is good at anything, it is accountability. We stay with our battle buddies and we for sure don't leave without anyone. Time to take a headcount. "Where's PFC Brice? Anyone seen Brice?" He probably overslept. So, they checked his room. Holy shit - he wasn't there.

"Who was with Brice last night?" I asked.

A couple of guys said they were with him, but when they turned around, he was gone. They assumed he went back to the hotel.

Oh no.

So, we waited. And waited longer. We waited for about two hours.

This bus was for other people, too, so we couldn't just keep them hostage because we lost a private.

No one had cellphones back then. We had them in Germany but there was no such thing as free roaming. Roaming in other countries was through-the-roof expensive. I don't even think the pre-paid burner phones, which most people had, even had the option of roaming. Our company commander was there, and he had a big ass work cellphone with roaming because the Army paid the bill. His face turned white with dread, and he fired up that phone to tell our battalion commander the news.

This is how I am guessing the conversation went... or in my mind at least:

Company Commander: Hello Ma'am. Captain X here.

Battalion Commander: Hello Captain X, how is Spain going?

Company Commander: Yeah... Spain...everything is muy bien, got a little sunburn, put some bocadillo's in my boca, drank some shitty Sangria, saw boobs on the beach, bought a Spanish hand fan that I will never use again in real life, oh and we lost PFC Brice somewhere unknown.

Oh no.

She gave him a plan. Leave two people there to figure out What

TF is going on and Where TF is PFC Brice. Leave them some money for a rental car, and they drive back when and if they find him.

Jesus Christ, we leave Spain without Brice. Poor young little Brice, where the hell was he? What if he was murdered? Could I ever forgive myself bringing him to his beach town of death? Or maybe he was kidnapped and on a ship to Morocco right now to be sold to some rich Moroccan that likes little white PFC's?

Two soldiers from his team stayed back. We waved to them from our bus filled with hangover farts and drove into the sunrise.

A few HOURS later, Brice came walking up all tired and confused. "Brice you are alive!!!" *HUGS* Then... "Brice where the fuck were you!?!? *ARM PUNCH*

So, it turns out, Brice did leave the club drunk, on his own, and walked aimlessly into the suburban area of the town. It started raining hard, so he picked a car in someone's driveway and crawled into the backseat to sleep. (What were the chances that the car was unlocked?!) Then the next morning/almost afternoon, a Spanish guy goes to his car to go to work, to find a sleeping US Army private first class in his backseat.

I can only imagine what Brice said to the Spanish guy before he ran out of his car to the hotel: "Ummm.... ¿Donde está la biblioteca??"

I learned the importance of accountability for all of your soldiers. I wasn't out that night with them, but look how easy it was to lose a soldier even when we were supposed to be all together. That was something I could never have happen again.

RAMPING UP AND TRAINING FOR WAR

There are certain pre-requisites that a unit must do before they are able to deploy downrange. Soldiers have to train in certain tasks, classes, weapons qualifications, vehicle standards, health checks, shots taken, etc. It is A LOT. It is a pain in the ass and extremely stressful for everyone. You look forward to just deploying so you have some downtime from all the incredible number of tasks you have to complete in the months ramping up for the deployment. Since we were Signal Corps, on top of the Army requirements, we had the communication requirements, to make sure that all of the equipment passed inspection and it was functioning.

So, when we went to the field that is what we were testing: did our comms work? Could we talk to each other? Could we pass data and provide internet to the area via our tactical satellites? Passing these tests made the unit able to deploy and support the war fighter. If our shit was broken it would make us look very BAD and it would be painful for the smart warrant officers to figure out where we were broke and how to fix it... fast!

Our first field exercise was to test all of our communications equipment. We were in the Grafenwoehr Training Area, in autumn, and staying in tents. Twenty-four hours a day we were testing the equipment in shifts. So, as the platoon leader, I had all of the shifts.

The first task was to set up the sleep tents, so we did just that. Everyone was hungry. There were no food options except MREs. I mentioned to my company commander that we could go get something hot. He said no. The rebel in me came out. We were the Renegades after all! I thought that my soldiers had SUCH a long, long day of convoys, waking up at zero-dark-thirty ... they deserved something hot. So, I got a partner in crime, Staff Sergeant Y, took a Humvee, and left camp to drive to the civilian side of Graf. Since the headquarters was not set up yet, there was no entry point, or trip tickets to be had, and we totally took advantage of this. This is also way before the time of cell phones, so once I was out of your line of sight, I was gone. The only way to get me to come back was tell someone, to tell someone, to tell me. We were out. We left with the lights off, totally incognito. Renegades.

There was a sign that read, "No tactical vehicles beyond this point." We looked at each other, literally wearing Kevlar helmets in a M998. "Ehh... we will be quick," we said and kept driving.

I had him drop me off at Taco Bell, and he went to park in the back where there were no streetlights. When I walked in, I looked like I emerged from the woods... holding my Kevlar from our Renegade Humvee in the parking lot (in all aspects) and I tried to pull myself together like everything was normal. Walking up, fixing my hair, maybe taking a twig or two out of it, and rubbing off a little camo from my face. "Good day sir, I would like to order two bags of your finest tacos."

I ran out of the Taco Bell with two big bags of tacos, like a robber who had just robbed a bank. It was actually pretty expensive, and I was a poor lieutenant. However, I was considered rich by my soldiers. Also, it was worth its weight in gold to just disobey my company commander. We went back to the camp, and I gave out the tacos to my hungry and appreciative platoon. I had one request: destroy and hide all evidence.

Then like a great comedy show ending, the next morning when I left our sleeping tent and whipped open the canvas door, what do I see right on the ground? A soft taco wrapper.

During this part of the training, we were out in the woods away from the garrison area of Grafenwoehr. We had our sleep tents, our operations tent, and the communication assemblages set up and sending/receiving comms. Well, we were trying to, at least. We were testing to see if our equipment would work when we deployed. If you get it to work, then you get certified that all will work, then it is packed up to get ready to deploy. Don't even breathe on it wrong because it might not work again when you need it to.

We were also planning who we were bringing for the deployment, who would be on each team, and who we would leave behind. On our single shelter switch (SSS) team, we had a very weak soldier. He was quiet, timid, shy, strange... whatever you want to call it. We knew we were going to a combat zone, so this was no time for niceties or sensitive feelings. We knew he had to be replaced. We were having the conversation outside of the OPS tent, and one of the female soldiers walked by with a huge bag on her back, carrying it with the other guys. She was a badass, and she was the same MOS we needed. I said, "Let's bring PFC C."

And we did bring her to Baghdad. For years later, and even today, I think about how her life changed so much because of that moment of her carrying her weight, literally, in front of us and I hope that it was the right decision for her.

In this field exercise, we were being tactical (for no reason) and pulling 360-degree fire guard. I guess this was good training for the times when our soldiers might pull guard duty downrange, but at this time we were on an Army base in the woods, and this only made the soldiers more tired and worn down. That's a lieutenant's perspective, but I am sure the sergeants major thought differently.

So, another training tidbit was that no one was allowed to shower. There were no showers. The only showers were back in the garrison area. The only soldiers who do shower during field problems are the cooks and dining staff because they are around and touching everyone's food. (This is a good rule.)

This is another training lesson that didn't make sense to me, because right when we got downrange, there were shower trailers everywhere. We should have been taught how to take combat showers (conserving water) and how to properly secure a weapon while showering. But, no, we were in the German woods smelling all funky together.

Around day two or three, I was approached by my platoon sergeant who said that the cooks were taking back female soldiers to the garrison area for showers if they wanted. I was surprised by this and didn't understand how this was fair. Being the only female in my platoon, and being the platoon leader, I told him that I won't shower until everyone else gets to shower. He replied with a nod, a hooah, and then walked away.

At the next platoon formation that day, he was out front putting out information to everyone and then starting almost preaching in his motivating way that, "Our LT said she isn't going to shower until all the Second to None Soldiers can!" Which got a big loud "HOOAH!" as a reply from the whole platoon. That made me feel good for sure, because I was doing what I felt was right, but it also sealed the deal for me that there was ZERO way my ass (literally) was going to shower during this field exercise.

How long was the field exercise you ask? Eight days! Eight long stinky, sticky, wash-yourself-with-baby-wipes days. The shower I took once we got back to the barracks of Grafenwoehr Training Area can only be described as: Glorious.

Being in this environment was the first time we had to shower with our co-workers. The field showers were like cubicles with no curtains. Having no shower curtains is so Army...it could make sense if we just had shower curtains, but there were none.

I remember reading some graffiti in the bathroom in the field. It said something like, "If all women in the military were lesbians, there would be no pregnancy or PT profiles, and all women would max the PT test." During these times, it was "don't ask, don't tell" and if you were caught having homosexual sex, or admitted that you were homosexual, you would be chaptered out of the Army. What a change from now, where we have a month dedicated to celebrating gay pride and we cut a rainbow cake in the dining facility for it.

While back in the garrison area, we stayed in old barracks with bunkbeds. However, the bunkbeds had a chain linked bottom and no mattresses. We didn't care. We were soldiers. Even though prisoners had better sleeping conditions in jail, we didn't complain. The only time we called bullshit was when I had to make a trip to the field grade officers building; we saw that not only did they have mattresses, they had layered them two or three thick.

Mattress-having sons of bitches.

I didn't say anything, and neither did my battle buddy when we got back to the gen pop. We didn't want to start a mattress riot. We were okay with not having them, but people would have been pissed knowing that the majors were sleeping on stacked mattresses like the princess and the pea.

. . .

MY PLATOON WAS in charge of running the range for the battalion. We put a lot of planning and hard work into doing this, with safety being the top priority. The day was very long and, at that point, we were all just exhausted. My driver and I had to go to the ammunitions handling area to drop off non-expended ammo after the range while everyone else went back to the barracks. We were driving the M998 Humvee on trails deep in the training area of Grafenwoehr. The signs all over the roads warned not to drive with the lights on and to wear our night vision goggles. We had them in the Humvee, and we complied. It was very hard to drive in the pitch-black night with the goggles on. It was the first time we ever had them on while driving. I was actually really nervous driving like that and was afraid that we were going to hit something.

We finished the tasks and came back to the open-bay barracks. Everyone was hanging out and having a relaxing time. This was before smart phones, so people were reading books, magazines, chatting, or listening to music on MP3 players. I remember walking in and thinking to myself, "Don't worry everyone, the hard shit is done." I looked over to see our company commander laying on his bed, reading the "The 7 Habits of Highly Effective People."

Are you kidding me?! We were out busting our asses for hours trying to get everything done from the range, with zero support from him. Then I walk in seeing him reading this book in the open bay. Was he reading that to impress everyone? Or was he really reading it for leadership tips? I wanted to give him Habit Number 8: fucking help your lieutenant.

OUR FINAL TRAINING for the sandy desert was in the middle of December, and we went back to the now snowy Grafenwoehr Training Area. Bavaria in the winter! I am sure the irony is not lost. Our signal equipment was gone already, so we were completing all the requirements that were needed to deploy - like live-fire exercise, shooting the long-ass M-16 out of the Humvee M998 - and other random training that we just went through the motions to complete. The live-fire exercise was basically teaching us how to do a drive-by-shooting. We were tired as hell, so we just put the weapon out of the window and shot into the darkness. One thing we did NOT do is what we actually needed, but it wasn't taught to us for some reason: How to react to an IED.

I remember one of the final days when we were all getting off buses and shuffling in the knee-deep snow into an open field, someone gave instruction about... something. The training topic is completely gone out of my memory because at the moment so many of us had a WTF moment and we were pretty much at our breaking points. The person was instructing about something, but you couldn't see or hear him because there was a horizontal snow blizzard flying into our faces (or whichever side of your body you chose to catch the violent snowflakes). We got back on the buses and pretty much all had the same sentiment: can we just deploy already... I am so fuckin' tired. We all have been working so hard every single day.

REACH FOR THE STARS

However, even during the times of working hard, I was excited to plan my promotion to first lieutenant. While sitting in the OPS tent of our field exercise back in Graf, I told my company commander that I asked Brigadier General Pollett to promote me. His reaction was one I didn't expect. He was sitting in a chair and then completely deflated like someone pulled the plug on a bouncy castle. Chest on his lap, he looked at me disgusted like I disappointed him. I thought I was doing something good. I also thought that I am my own person and have the right to ask who I want to promote me.

Brigadier General Pollett was the first general I had so much interaction with. I thought this was a normal thing that happens in the Army, but as time went on, I learned that it wasn't, and he was very special. I asked him to promote me to first lieutenant and he accepted. It was not common for a second lieutenant to ask a brigadier general to promote them, but I always "reached for the stars" (no pun intended).

We had my promotion at the dining facility right after the Thanksgiving lunch. Almost all of my soldiers were there. Brigadier Pollett and Sergeant First Class Penton, my platoon sergeant, were the ones to pin me on both shoulders of my dress blue uniform. This was my first promotion ever, and it was a nice ceremony. I had to fill out an information card about myself prior to the promotion for the general. This included where I was from as well as where I went to school, my hobbies, and my goals. He memorized my entire bio; when he got in front of everyone he didn't need the card. I was completely impressed and felt special.

He came around the motor pool often and unannounced. Again, I thought this was common practice, but it really wasn't. You could be hauling items in the connex or doing maintenance checks on a vehicle to turn around and see Brigadier General Pollett standing there with his Kevlar on! When he came around to meet the soldiers he would ask them questions about themselves, and he would ask to see family photos, if they had any in their wallets. If a soldier didn't have a family photo in his or her wallet, he would tell them that he would check next time ... and he most definitely did!

I had no husband, no kids, and no wallet for that matter... so I was off the hook. I guess I could have produced a photo of my evil cat Minnie that I had at home.

Auf Wiedersehn Germany

I loved Germany so far. Of course, I was bummed to leave beautiful Germany and go to Iraq. I loved my apartment, my shitty car, and the freedom to travel and see new beautiful places. I was in a constant state of awe how gorgeous life was around me. Coming from NYC and dirty Jersey, Germany was like walking through a beautiful dream. I was sad to leave.

Each week we were told we were going, then it would change. Nothing feels better than canceled plans. Well actually there is something... a canceled deployment to war. Okay not canceled, just postponed a week, but it meant we could go do something. Back then I stayed away from desserts and pastries, but on the newly found free weekend I went to the gorgeous city of Heidelberg and bought a delicious fancy pastry. It was a picture-perfect gorgeous day and I walked high above the city up to the Heidelberg castle to the edge of the gardens that look out on the Neckar River. I sat there and took in the view while eating this normally forbidden treat. I got the treat and did this trip because I thought that maybe this could be my last time in Germany. I might as well enjoy it and soak up the feeling of being there. I really thought there was a possibility that I could die in Iraq and this would be my last time there. I was a young twenty-four-year-old girl sitting at a tourist attraction in a crowd of other tourists, but little did people around me know how sad and dark my little trip up there really was.

Chapter 4

1Lt

We had some delay with the original deployment timeline because we were supposed to enter through Türkiye, but it was voted down. We were on a hold status, but when there was a clear date for our deployment, they let the unit go on block leave right before the deployment. Block leave meant that everyone was allowed to go on leave at that time. Normally it would be staggered to make sure that someone was still there and the unit was still functioning.

I went home to New Jersey for a week or so to visit family. I tried to play it off like it wasn't a big deal so my Mom wouldn't worry - how I was only in communications, and we didn't do so much dangerous stuff. (All not true.)

While on leave in New Jersey, I stayed at a nearby beautiful hotel, which was designed to look like a French chateau, in New York state right on the border of New Jersey. In the early morning, the hotel staff put a newspaper under each guest room door. The front of the 15 December 2003 newspaper read, "We Got Him!" And there was a photo of Saddam Hussein all scruffy with a beard and long hair. He was in an eight-foot hole under a hut?! That's not what I was expecting... BUT... the big question was: Does this mean that we don't have to go now?!?

Should I call my unit? Have plans changed? I mean we were looking for him because he had weapons of mass destruction, but it seems he was in a hole under a hut and not in some underground high-tech evil lair with all WMDs around him... so can we not deploy now?

I really thought that. To me it was clear. Or worst case, we go there and don't have to stay for a year. Just show up and like, yeah, we are here because we already booked our plane tickets, so we will stay for a little bit to close everything up then go back home.

Well clearly, I was wrong. Hopeful... but very wrong.

AFTER BLOCK LEAVE, everybody-lodi-dodi reports back and is accounted for. Usually what would happen was a commander would want to see who had *too* much fun on block leave and conduct a 100 percent urinalysis. This might even be at the first formation after block leave. Everyone with no exception goes upstairs and lines up in the hallway ... and starts drinking water because no one is leaving the area until everyone can pee in a cup while someone watches.

Our company didn't have anyone "piss hot" (meaning test positive for drugs) but I heard a story about a lieutenant from another company who did. She went into the stall with the female NCO who was administering the urinalysis, and the lieutenant asked the sergeant first class to pee for her. The sergeant first class declined and went immediately to the company commander. I'm not sure how stupid the lieutenant was to smoke weed on block leave before a deployment... but it is also even more stupid and selfish to ask an NCO to put her entire career on the line to cover for you.

We departed Taylor Barracks in Mannheim, Germany at 2200 on the thirteenth of January 2004 with rucks on our backs and unknowing anticipation on our minds. The energy in the air was unreal. Some people were excited but with the feeling of the unknown. There were families scattered everywhere, out in the field and parking lot, and crying... it was a moment where a big new chapter in our lives was starting, and everyone knew it. Walking around, I would see a bunch of soldiers and they looked excited like we are going on a big trip (well we were) and I would try to smile and motivate them; then I would turn around and see a family with five kids who were all crying, and I would stop smiling and put my somber face on.

One soldier arrived late! They were ready to write him up for being AWOL and also deserting a deployment, and he comes in late from his German girlfriend's house! He wasn't in my company, and I was glad for that because I would be freaking out. I walked by his room while he and his leadership were frantically throwing in items into his duffle bag.

My first platoon sergeant, Sergeant First Class Penton, came by to see me off, which meant a lot to me. He was being assigned to another unit, and SFC E was talking over the platoon. So, I walked through the parking lot of crying families, and met him by the road near his car. He had a bag for me, with some items that helped him with his deployment in Israel, and he was giving them to me. "Here take this, put this in your bag," he said.

Then he gave me some anti-diarrhea medicine and looked me in the eyes and said, "LT, listen to me, take this just in case..." We both kind of looked serious at each other, then laughed about it. It felt really good that he cared about me so much.

Throughout the rest of the deployment when I saw the box of anti-diarrhea meds in my stuff I always laughed, thinking about how serious he looked at me when he gave it to me. So let this be a lesson to you, if you are leaving for a distant land with minimal indoor plumbing, and your mentor, friend, or partner doesn't give you anti-diarrhea medicine...they don't *really* care about you.

The long convoy of German buses took us to the Rhein Main Air

Base staging facility. Some units stayed there for a long time, but we only stayed there for a few hours. They had ping pong, pool tables, food, TV's, and USO goods with letters from high school kids who donated everything. I read one from a sophomore from Nebraska who liked to play football and go hunting. Also, I thought it was pretty funny they edited his letter with black marker, but you could still see that it said, "I hope you guys kick some fucking ass over there." Thanks dude from Nebraska, so do I.

The staging area was extremely organized. There were drinks and snacks out, reading material, and rows and rows of made-up bunk beds. When we walked in, they had a soldier literally at every corner telling you when you had to turn leading to your bunk. It was definitely overstaffed, but I think it worked out great. So, I looked around and ate a Jimmy Dean snack they offered.

I decided to take a nap for two hours until we had our formation at 0420. While I was getting out of my bunk and just waking up, I saw right in front of me Brigadier General Pollett, the 5th Signal commander! (I told you how he just shows up!) He came by my bunk and said hello to me and the rest of the soldiers. I thought that was so awesome and I was so proud that he was the one who pinned me for first lieutenant.

The flight was on a nice commercial aircraft and lasted only about four hours. Actually, we would have liked the longest possible flight because then we could sleep more. When we arrived in Kuwait, it was a huge change from beautiful Germany of course. There was one rustic-looking dirt road we flew over right before the airport. I leaned over to a fellow Jersey-ite, "Look it is the NJ Turnpike!" Because when you fly into Newark, all you see is the lights of the New Jersey Turnpike below you.

We got on another bus and rode to some big area to sort rucks and duffle bags. Everything was totally surreal. We sat on the bus for a while until the logistics folks figured out the plan. We sat and looked out the windows, as if we landed on Mars. It might as well have been Mars. Nothing but sand as far as we could see, and there were Kuwaitis working in the area and driving buses. This was only a little over two years after 9-11. I had never seen someone in real life dressed in middle eastern clothing, other than on TV and the news with videos of terrorists and ISIS. And there we were, and they were driving our busses and sorting our bags. Of course, I am not saying they were terrorists, but up to this point, that was what I had seen portrayed in the media.

We walked through mud and got on another bus that had lots of dried mud on the floor and a driver who also had dried mud on his boots. Did I make my point? The bus driver was the first Kuwait citizen I was in close proximity with ever. I wonder what he thought of us ... with weapons ... on his bus. He drove us on a main paved highway to take us to our final camp.

Almost everyone fell asleep on the bus, including me. Then the bus driver turned onto on a dirt road in the desert, which was marked by two Kuwaiti soldiers in a Humvee. I knew they were Kuwaiti Army because one of the soldiers had a huge beard! So, since we turned onto this dirt trail in the desert, I thought that the camp we are going to must be very close. But it wasn't. We drove for about two hours through the desert. We were practically off roading with buses. We got stuck in the mud and had to get out and push. It was pretty funny actually... after the fact, not during.

So, we arrived at Camp Udari. At first looking at it through the bus windows, we all thought, "God, this place really sucks." But once we drove to where we are staying, it wasn't that bad after all. And perhaps once the sun came up, it could be quite pleasant. They said the weather during the day was like a cool sunny day on the beach, and this was during their winter.

It seemed that Camp Udari was made of mud piles, plywood, and sand out as far as the eye could see. We lived in a tent city which, from a distance, looked like the circus is in town. (No pun intended.)

All of the soldiers, regardless of gender, slept on cots in a huge tent together. I would say there was more than 100 people in my tent. They tried to do a lot for the soldiers to improve the conditions and morale. There was a PX which had pretty much everything we could need or want. Also, they had some items that would go real fast because of popularity. Right away, I bought a great desert fold-up chair! This was awesome because it gave me a place to sit besides my cot, which was not so comfortable and had zero back support. It might have been the best eight dollars I had ever spent up to that point.

The first couple of days we were not setting up our signal equipment yet, so all of the soldiers were hanging out in the tents watching movies. It was actually a lot fun and nice to not be so busy like we had been the past few months. One soldier bought a TV and set up a little movie theater. Then everyone brought their cool eight-dollar chairs with them to watch the movie. If we didn't go to the movie theater set up, then we watched movies on laptops. During the day, soldiers also read books, played video games, and played cards... spades mostly. When shopping at the PX on Udari, the average soldier's shopping cart consisted of a couple cans of dip, Mountain Dew, Black and Milds, chips, gum, and sunflower seeds. These were the staples for the movie watching or card (spades) playing.

ONE DAY, I was sitting in the tent listening to the White Album by the Beatles on my headphones. The song 'Black Bird' came on, and I got sentimental and deep with the lyrics.

"Take these broken wings and learn to fly. All your life. You were only waiting for this moment to arise.

"Blackbird fly, blackbird fly, Into the light of the dark black night."

It had such great meaning to me at that moment. Maybe the Mountain Dew was getting to me, but I saw my soldiers as the black birds. "Take these broken wings and learn to fly." No matter their backgrounds or cultures, where in the States they came from, the type of families they came from, the problems they had, their ages... we were all soldiers now, learning to fly.

That day my commander told me that my platoon was chosen to move north to Baghdad City in the next week, and this was going to be our permanent station for the rest of the deployment. This was what we had been training to do and we were ready. This was our "moment to arise."

The unknown, the danger of the three-day convoy we had ahead of us, the thought of our loved ones at home, not knowing how long we will be there... we were traveling "into the light night of the dark black night."

PORT-A-JOHNS

PORT-A-JOHNS, Port-a-Shitter, Port-a-Lu (UK), call if what you want, but unfortunately while downrange, it is often a topic of daily conversation - normally in the form of a complaint. The graffiti on the inside walls was out of control in Kuwait and Iraq. I don't know what possesses people and why they have the urge to mark up the inside. You don't do that at home!

Disclaimer for myself: I never once wrote on the walls, nor had the urge. But like it or not, it was entertaining to read.

The photos were sometimes funny. There were some good artists. However, some made me wonder if they should be seen by a psychiatrist for a mental evaluation. This was before the personal cell phone; I realized later that they were just super bored. There would be such graphic drawings on the wall that were literally sharpie marker porn. They had to start posting guards outside the port-a-potty and they had to check inside after someone was done to see if they drew on the wall. Shitty duty. Double pun intended.

You could find personal gossip and information about certain soldiers with their names, which I didn't think was cool. Was that like an early form of cyber bullying? It was on the walls of the port-a-potty for all to see... but I guess it is called port-a-potty bullying. You could walk out of it and say, "Did you hear about sergeant so-and-so? I heard blank blank... yeah, I read it on the port-a-potty."

I wrote down some interesting sayings written on the walls:

"Welcome to Kuwait... Now look busy."

"Got Food? Water? Fuel? Bullets? ... Got Jesus?"

"If life is one big joke... I don't get it."

"Respect My Auth-or-ity!"

"Victory Base: Our home away from home."

"Jesus is my co-pilot on the way up to Northern Iraq."

"BOHICA... Bend over, here it comes again."

SPEAKING OF PORT-A-JOHNS, I got hit on by one of the workers once and it was so incredibly awkward. He was literally sucking the contents out of the port-a-john with a huge hose that was attached to the "honey truck." The "honey truck" smelled soooo bad and it made a clear line of stench that spread down the sand through the camp like a stench-laser. If you weren't in the line of the stench-laser you were safe, but if you were in the line you HAD to hold your breath. There was no other option. Either hold your breath or you would literally gag.

So, you get my point about how much this thing smelled?? And the man in the epicenter of it all, (somehow still surviving) actively sucking up everything, looks over to me and says, "Hey baby, what's your name?"

My mouth was sealed shut by my lips like a tomb, and I shut off all breathing through my nose, because I was walking through the stench-laser. Still trying to be nice because I was a representative of my country, I didn't want to him to feel like shit (no pun intended) and that I was too "good" to talk to the port-a-john guy.

But I physically could not speak. Speaking meant the honey truck fumes would enter my nasal passage or mouth, and I would be gagging in front of him. So, I just grinned, nodded my head, and hummed to him as I walked by fast. "Humm mmm mmmhummmmm hummmhumm."

EVERY SOLDIER'S NIGHTMARE...

Setting up in Kuwait

A good way to describe how the building/rooms in Kuwait smelled is to image that someone mopped the floor with armpit sweat. Also, you can smell the heat. Yes, heat has a smell... and it isn't pretty. When you are by a body of water, it is so humid you feel like you can cut the humidity in the air. Or swim in it. Breathing deep does not feel enjoyable. You feel wet all the time... your clothes feel damp, and you smell like the heat.

Out in the desert there is a different heat because it is heavy and dry. It just weighs you down and you feel like you reach a point where you can't be any hotter and you just deal with it. You don't sweat because it evaporates too fast.

THE ROCKY ROAD TO BAGHDAD

I was selected as the convoy commander of our group that was headed north to Baghdad. I found out that there was another signal unit at the camp that would be headed to the same place we were going, but they were leaving a few days sooner. They had all the training and all the weapon power that we would need to be successful, because let's be honest, my unit was lacking in both those aspects. We had been doing a lot of training, but we never did any convoy training or IED reaction training. I was so determined to make sure that we connected with the unit and were embedded into their convoy. I felt like this was the only way that we would be successful and safe going to Baghdad.

The leadership was not being helpful or proactive to link us up in the convoy. Everyone was setting up their own OPS tents. I walked over to other unit, which was from Fort Hood, and met one of the platoon leaders. We made a plan of how and when my platoon could link up to join them.

I was very frustrated at this point because no one else senior to me had a sense of urgency about us getting into the convoy. I felt like if we went on the convoy by ourselves, it would be dangerous and we be an easy target for the terrorists. We had all soft vehicles and two M249 machine guns but no 50-cal machine gun, smoke, or working radios.

We also had no training so far on how to react to an IED. The only training I received was from a Special Forces soldier I had met in the Kuwait dining facility; he gave me a quick lesson using salt and pepper shakers, with the IEDs being the sugar packets. I told him I was scared to go on the roads to Iraq. He told me that it is normal to be scared... especially in our situation! He said, "I'm afraid of the person who isn't afraid to be going on this convoy, because that is the person who will fuck around and get us hurt."

We had one day to get our vehicles ready to go on the convoy. We had to cover up all bumper numbers, so the terrorist didn't target the leaders' vehicles. If it says A6 you are the company commander, and A7 is the first sergeant. Since I was the second platoon leader, I was A206. We also had to scrape our names off the windshields of the Humvees. To harden our canvas soft Humvees, we had to go to the scrap yard and get some plywood and old flooring to protect our doors!! We put them inside the doors and also above our heads, so if a grenade fell down, it would at least bounce off the plywood. Unbelievable...

The next day, my soldiers let me drive the Humvee. We did this because if they got hit and couldn't drive, I would have to drive. So, I needed some sort of familiarization with the Humvee. Officers don't ever drive themselves, (also no license) so this was a first time for me, and it was awesome!

· · ·

WITH A LOT of determination and tons of hard work from the soldiers, we pulled it off! We met up with the other unit we were going to convoy with on the edge of Udari and lined up in one big convoy broken up into groups. We were embedded in their convoy, so even though we were driving by ourselves, we knew that there was another group ahead of us and behind us.

Since our single channel ground and airborne radio systems (SINCGARS) in the vehicles did not work, the unit we convoyed with let us use four hand-held radios. I used them to control the convoy, pass information about the surroundings, and keep accountability of the soldiers. I put the radios in the lead, middle, and rear vehicles. When you are in this type of tactical situation, you realize how important communication is for a mission.

A lot of people came out to the convoy line up to say goodbye to us. The chaplain came to me, started crying, and told me he didn't want me to die. (Seriously) Other companies came and gave us some of the items we were lacking or could use more of, like extra fuel cans, MREs, flares, and drinking water. I thought this was pretty awesome of them.

We had a 50 cal, but we didn't have a mount for it, and we had no ammo. It was a ridiculous situation, but we did the best what we could... and also added a little humor. My soldiers hung the 50 cal from the back of the deuce-and-a-half truck with some engineering tape. When we drove, it would swing from side to side, BUT it kind of looked like we had that kind of firepower... from a distance... if the terrorist had bad vision... but it was better than nothing! We looked like something out of Mad Max. To be funny, they took some plywood and wrote, "Ice Cream $1" on the side.

DURING A CONVOY you keep about 100 meters distance from the vehicle in front of you; this is so you have time to react. Also, if you get hit with an IED, less people are in the kill zone. We left our safe bubble, Camp Udari, and departed toward the border between Iraq and Kuwait to Nav Star, which was a two-hour Humvee convoy drive from Udari. The pitstops along the roads were called military supply routes (MSRs) which was where we received fuel, intel briefings, and maps to our next destinations. That night at Nav Star, we set-up our cots right next to our lined-up vehicles and slept a few hours. Our wake-up was at 0100, so we are not talking about getting a very good or long sleep.

The places we stopped at were pretty surreal. The perimeter was fenced, there were many rows of port-a-potties, so many stray dogs running around ALL over the place, and tons of trash on the ground around the vehicles. There were layers and layers of trash in the sand and gravel from all of the different groups who came through. If you were lucky they would have a room you could go in and they would serve some warm soup. I never saw this, but my soldiers told me about it, and they thought it was pretty awesome.

When we woke-up to depart, it was freezing cold outside! I couldn't believe the climate change, from hot to cold, in the desert. So, we got in our vehicles and rolled out to cross the border into Iraq. They call it "crossing the berm" because of berm barriers that were constructed as a huge wall. When we crossed the second berm, our orders were to lock a twenty-round magazine in each of our weapons and put the selector switch to semi. Everything was getting very real.

We came up to the barrier. It looked like an ugly New Jersey Turnpike toll. There were signs that read, "You are now crossing into a war zone," and "Watch out for children in the road."

Jesus! Children in the road! What a freaking nightmare. It is hard to understand how the enemy would strap a bomb to a child and send them into a convoy. I heard about people being very affected years later after the deployment, about how the terrorists would tie children's hands together in a line and send them to spread across the road so we would have to stop the convoy. Then if and when the convoy stopped, they would attack with IEDs and rockets. Again, an absolutely nightmare of a situation, because our rules of engagement were to stop for nothing in our path.

The minute we passed the sign, "You are now in Iraq," an unbelievable rush of adrenaline ran through me and everyone in my Humvee! It was surreal. What the hell are we doing here? Are we really in Iraq?!? How the hell did I get here? I was in Germany a week ago, got on a plane… hung out with my friends in Kuwait… drove a few hours… and now we just entered the pits of hell and a war zone.

My knees were shaking like cartoon characters do when they are scared. It really wasn't fear that was taking over me; it was the rush of being in a real-world mission and having my loaded M16 hanging out the window. Also, my whole body was shaking because it was absolutely freezing, and we had the windows open going about fifty-five miles per hour in a Humvee with no heat.

My driver was Specialist J, and he was my driver for a reason. He was the sharpest and most mature specialist in my platoon. I chose him to be by my side and help me as the convoy commander. In the back I had also a very smart satellite communication systems operator, Specialist C. I selected him to drive with us because he would help us navigate the convoy. He was incredible. He used the ridiculously HUGE fold-out terrain map that our operations officer gave us; he tracked the terrain changes in the landscape, and he knew where we were at all times. It was truly incredible that they gave us this massive map and told me to drive to Baghdad. Luckily, at the pitstops, they gave us photocopied hand drawn maps of the MSRs and the upcoming pitstops. So, with Specialist C, those handouts, and a path of spray-painted plywood signs with arrows, we were set.

AT ANOTHER PITSTOP, since I was the convoy commander, I had to go in and receive the briefings. All of the other elements of the entire convoy were there, so the signal unit from Fort Hood, that basically saved us, was with us as well. There was a very nice captain with the unit, and he asked me, "Where is your company commander?"

"He is sitting in his vehicle," I said as I pointed to the lined-up vehicles outside.

So, he walked out to the line of vehicles and brought my company commander inside to the brief with me for support. I appreciated this.

DURING THE BRIEFING they told us that it was too early, and the IED clearing team had not been out yet. (Oh, fantastic.) Then they told us to not tell the soldiers because it would make them nervous. Well, that is not how I operate. I went back to my platoon, and the first thing out of my mouth was, "Hey, the IED clearing teams have not been out yet, so we have to be EXTRA careful and diligent."

WE CONTINUED to drive in the dark for many, many hours. We were scanning the darkness looking for anything that might be an IED or some suspicious person in the sand dunes. Piles of trash, soda cans with wires, roadkill, water bottles, MRE bags... these could all be roadside bombs / IEDs. Also, anyone holding a cell phone or some kind of detonation device could be on the side of the road. When the sun came up, it became clearer what we were staring at in the dark. It was just miles and miles of sand and garbage. There is no other way to describe it.

Around 0800 we started to see cars on the road. The cars were mostly total junkers! Pieces of shit that would never even pass even a lenient New Jersey inspection. They didn't have lights on the back of the vehicles, and the front grills were all bashed in. Any car that came up on my convoy got radioed in from the lead vehicle and each person followed the car with the muzzles of our rifles. There were too many suicide car bombings recently, and I would be damned if we were going to let it happen to us.

Back at Camp Udari, a Special Forces soldier told me, "You have to keep a serious face and look them right in the eyes and let them know you mean business, so they won't mess with you." He told me that Iraqis who were attacking from private vehicles were mostly hoodlums and didn't want to die either. He said they were waiting for the soldier who was sleeping and not paying attention or with his Kevlar off his head. God, what good advice this was. So, we had to play the game. And that is what I told my soldiers before we left.

My soldiers were signalers by trade, but still confident to pull the trigger if necessary. One car was riding along the side of our convoy, going from vehicle to vehicle. I radioed to the trail vehicle, which was our gun truck, and he told me he had eyes on him. Then after a few minutes he radioed me back saying that the car was driving alongside our gun truck and keeping pace with the convoy. This news put me on even higher alert, and made me nervous. Then the gunner on top with the M249 pointed the squad automatic weapon (SAW) right down onto them. So... that shook them off us.

The day went on, and we drove deeper into the north and started coming across more Iraqis in cars. Suddenly, the road split and there were two options: 1. the rural road or 2. the city road. Soldiers at one of the checkpoints told us that we were cleared to take the rural road. What a relief! We drove along this road for about three hours, and I will never ever forget those three hours for the rest of my life.

Just picture miles and miles of white sand dunes to your left and to your right. Along the road there were little clay huts, and farther back there were larger clay huts. I have never seen this before in my life, and I still don't understand where the people came from and what they were doing. They were just standing there a few meters from the road, or just squatting in the sand. Men were in one group, and women in another group. They all wore skirts with wraps around their bodies. I felt like I was in a movie. There were YOUNG children along the road by themselves without a hut to be seen for miles. These people were so poor, they were on another level poor. There was nothing around but the hut and then us driving through. It must have been very exciting for them to see us. They were engulfed in dust that our convoy was making but they still sat there just watching us.

WE CAME across a few randomly placed people, and they were so happy to see our convoy. They were waving like crazy, they were dancing so freely, they gave us the peace sign, thumbs up, and most of all blowing us kisses. The young children all came out to see us, and they all were blowing us kisses, too. A young boy had his two-year-old sister right next to him and was putting his hand on his sister's mouth and blowing a kiss for her. A man wrapped in a colorful robe looked right at my vehicle, pressed his hand to his heart with such feeling, and then put his hand out from his heart right to us. Wow.

I have never felt like that before. After all the news and what the US Army did in Iraq, after all the real-world mission training, after packing up everything for the deployment - it all started to seem worthwhile and very real. They weren't thanking me personally; they were thanking the entire US Army for what we had done and that we captured Saddam Hussein. It was such an indescribable feeling of true pride.

IRAQ

Arrival

WE ARRIVED at Victory Base very early in the morning and entered through one of the main gates. I didn't know it then, but I would never see that main gate again. I was surprised that it was so close to the city streets and buildings. There were people in the street walking up to the convoy trying to sell soldiers bottles of some kind of liquid. It looked like used water bottles refilled with juice or iced tea. Also, I heard they were trying to sell alcohol. It was extremely shady and completely dangerous in my opinion to buy these drinks.

The roads were dusty and very wide across. We were a huge convoy of military trucks and there were still people walking in the road with plenty of room to set up their mobile "shops in a box." I can't blame them for trying to make money, but I don't know anyone who had cash on them or was willing to buy anything.

The unit that we were replacing met us at the gate. We first came in contact with someone on a satellite phone. They were really happy to see us because we were their ticket out of there. In signal, you do a relief in place (RIP) where you strategically swap out their equipment with your equipment without any break in the communications. Once that is established and the communication is running well with no problems for a few days, then the unit can redeploy back home.

Victory Base was always buzzing with energy, people, helicopters, and vehicles. We were there during a strange time when the first wave was leaving and the second wave was setting up how they were going to operate. The palace was empty because they were going to renovate it and refill the area. People were swapping sleeping locations, and everything seemed very temporary. A temporary mailroom, temporary PX, temporary unit headquarters, etc.

IT WAS TOTALLY UNEXPECTED, but when I arrived, the first few people I met were from my officers basic course back in Fort Gordon, Georgia. It is a small world in the Army, but even smaller world in the Signal Corps.

Two of these officers were female lieutenants I knew from OBC. They looked tired and like they had seen some terrible things in the past year. I felt kind of guilty that they were deployed right after OBC while I was enjoying beautiful Germany the whole time. They were also stationed in Germany, in Darmstadt, which was about forty minutes away from Mannheim, but were quickly deployed away after arrival.

One of the officers met with me and my company commander to talk about starting the RIP process right away. Right away? We hadn't even put our things down yet. We just drove for three days, we were really tired, we were dirty, and we felt worn down. I suggested that we start the next day to let everyone shower and sleep first. For the benefit of the soldiers but also for safety reasons. The major said he still wanted to start right away, and my CO didn't argue with him. So, we started the RIP immediately.

While in the RIP I talked to one of the lieutenants that I knew from OBC. She told me that one of her soldiers died by suicide in one of the signal shelters on his night shift. He shot himself with his own rifle, and when she opened the door, he was lying there, and the floor of the shelter was filled with blood. Absolutely horrible. It also scared me to think that it could happen to one of my soldiers. I promised myself to do everything I possibly could to prevent that from happening.

I always tried to be positive and find the best in a bad situation as a leader. I hoped that this would in some way help the overall morale of the team. They knew I genuinely cared about their wellbeing. I did everything that I could to make sure that we all went back to our families at the end of the deployment.

Our new home sweet home: Victory Base in Baghdad, Iraq. I felt like we were lucky compared to others deployed to different areas in Iraq. The base itself was "nice" because it was made originally as a resort location for Saddam; however, it was never completely finished. There were palm trees everywhere, and the buildings were really beautiful inside and out with colorful designs made with beautiful painted tiles. We walked everywhere we went, exploring new things to see and ways to get around the sort of maze. You wouldn't really see the same people twice since there were so many coming and going all of the time. We had no cell phones, so we had hard times to be places to meet someone for a shift change or meeting; but other than that, were basically free to roam. Of course, soldiers were on a tighter schedule with their NCOs, but I had a lot of freedom.

Our living conditions were pretty comfortable, and we lived in trailers and not tents. They called them CHUs - containerized housing units. No one called it a containerized housing unit, and most people didn't even know what it really stands for... it was just a CHU. (Pronounced "chew")

The CHUs were nice compared to tent living. We were in a big CHU trailer park. The room inside was like a small dorm room. We also had heat and AC inside each room. At the end of each row of CHUs there were toilet trailers and shower trailers. Next to them were concrete shelters to go in if there were mortar or rocket attacks. These CHUs were new to the base, and the other soldiers who had been there for a year before us lived in tents.

ONCE ALL THE comms were set up, after the first couple of days, most people in the platoon got SO sick. It was known as the 'Iraqi crud'. When you said you got the Iraqi crud, you would say it like you had the flu, or bronchitis, like it was a real medical condition... and everyone knew exactly what was wrong with you. People thought that it was the new germs and that your body was trying to get used to it and fight it. This could be true. If you did catch the crud you didn't make a big deal about it because there was so much real-world conflict all around soldiers getting killed or severely wounded. There was no way you would walk in the stand-up hospital there to tell them you have the crud because you felt that it would take attention away from soldiers who were coming in injured. The crud was similar to a cold with severe congestion. I got the crud, and it hit me like a ton of bricks. I don't think I have ever been that sick with those kind of symptoms in my adult life. My nose was completely blocked with congestion, and my head and chest hurt so much from it. We didn't have any medicine, or any cough syrup... we just delt with it and carried on. We were soldiers.

The other female soldier in my platoon got it really bad as well. She went to the hospital because probably someone told her to go. Since I heard she was in there, I went to go visit her. Inside they had one big room with medical leather-type beds with curtains separating the beds into small "rooms." When I went in the tiny, curtained room she told me, "Screw this I am leaving... I can't be in here with everyone that has real problems." The hospital was all for triage and emergencies. People in the other rooms were shot and wounded during convoys or while outside of the base on patrol. She didn't feel right taking up a bed, so we left to go suffer our crud together.

On one of the strongest crud days, I slept-in later than usual. I mean, I would wake-up to go sit in the OPS areas and be there until 0200 or something, so it wasn't like I was not doing my eight hours of taxpayers' money. But when I showed-up, my company commander came at me all serious asking where I had been because I was late. I looked at him completely shocked that he couldn't tell I looked like absolute shit and my nose sounded like there was a clothes pin on it. With the most nasal sounding answer ever I said, "I have the crud...I was sleeping." The overall feeling was that he did not give a shit about my health and only cared about where I was that morning. Showing up at 1000 hours ... wow.

My new desk space was about a twenty minute walk from the CHUs. Not a bad commute to work. The unit we were supporting gave us a really interesting "platoon headquarters" that was in the middle of a big clearing under a huge antenna with a flashing beacon on it. So basically, the beacon was a mortar and rocket targeting practice area for the enemy, and we were directly underneath it.

It was a complicated situation because it was my platoon that was there, but the company commander also came with us to Baghdad. So, we decided that he would stay in the trailer in the field, and I luckily found a spot closer to where my soldiers were located to call my office.

I took great pride in being their platoon leader, and my soldiers knew this. I was always around, and I showed that I genuinely cared about them and their well-being. In the signal world it is important that officers don't try to take over the technical aspects of the mission. I was the bridge to the company commander and the battalion commander. I was the voice of the platoon. If we needed something or had a problem, they would tell me, and I would go fight for it.

I learned to be there and see if they needed my help, but didn't stay too long. There is a fine line between being a part of the team and still being the platoon leader. Soldiers are not going to act like they normally do when officers are around. So, it was okay to be in the shelter area and stop by to see how they are doing and their set-up, but it's not okay to be there all-night playing dominos like one of the guys. There has to be a clear separation, because when you need to stand up and be the leader, you sometimes won't be everyone's friend. If they already saw you as their equal and a real buddy, then you stand in front of the platoon with some info/news that no one wants to hear (which most of the time came from higher), you look like a fake and a traitor. "Oh, I thought they were cool..." There must be a careful balance as a young officer.

I learned to always be there when times were hard, like when the comms were down and everyone was out in the rain trying to get the generator back up. That was not the time to be in the OPS tent. I had to be out there and be present in 'the suck.' We were a team and my platoon needed to know that when things get hard and rough, I would be right by their side.

Our platoon's data package was spread out in a few spots in the middle of the base, right next to the big palace. The single shelter switch (SSS) was right behind what we called building 1, which was the embassy at some point, and it was filled with a lot of people and desks. The SSS was the backbone of the data package and it controlled everything. You could also attach other types of signal equipment to the SSS to provide services. The soldiers in there were very smart and knew the ins and outs. As a platoon leader, I was not the one pushing any buttons. (Good thing for that.) A lot of people would visit from other units, mostly the technical warrant officers. They wanted to see what was going on and also see how they could expand services off it. We had an awesome team of Renegades in there, and they pulled a twenty-four-hour shift monitoring the SSS. And of course, one knowledgeable and 'takes no shit' NCO was in charge of it all.

The SSS was connected to a sixteen-foot satellite dish which was about fifty yards away right next to the river that ran through the base, which then surrounded the palace as lake. This dish was connected to and pulling down services from a huge satellite distant end in England. We had at least two connections, so if one went down the other would pick it back up.

THIS SATELLITE TEAM of soldiers were so smart! Their NCO was intelligent but also so funny, and a great soldier who did his job and also made it a pleasant time for the soldiers there. They set up a tent in front of the shelter, with a door and a walkway. They would play dominos and even did some fishing in the river with the rods they ordered online. I caught my first, only, and last fish there with them... and then I threw it back in. They knew how to make a good time out of a bad situation.

THESE SIGNAL ASSEMBLAGES were all connected to data boxes and this funky piece of new equipment (at the time) called a Promina, which tied everything together. This was located in an actual building. The building was one of the houses built there for Saddam Hussein's vacation village. The whole area was built with a resort and vacation feeling to it in the architecture; there were palm trees, zoo animal cages, and a palace in the middle of the lake. The building I was going to work in was a small, one-level building made of concrete. There was a living space in front, perhaps it was supposed to be a big bedroom to the right, with a smaller room to the left, and then *drum roll* a bathroom with shower. The walls in the whole place had decorations made with colorful tiles. It looked fancy, but also looked like it wasn't done yet. The invasion halted the construction in the whole area.

There was a contractor company called ITT in the building, and they were also in charge of providing services there. We somehow worked with them hand-in-hand and especially with the data Promina part. They had about five contractors there, making A LOT of money to basically ensure that Promina was working. They would tell me how much money they made to be there, and it would blow my mind. My data soldiers were there to set it up and manage the equipment also, but there were all those contractors in charge of the equipment as well. Then there was me, the lieutenant. I was in charge, yet pushed aside, because the five contractors were being paid a lot of money to be there and do their job. Regardless, I was still responsible for sending status reports back to the battalion commander in Kuwait.

In addition to the data contractors, we had the heads of the team in the building. They were responsible for a lot more than that building and also for other locations around base. The head contractor there was Shane, and we clicked right away when we met. He was really nice to me, and I think he could tell I was like a stepchild and that my commander wasn't very fond of me, so he said I could have a desk in the office. This was amazing to have! I felt incredibly grateful and lucky that Shane took me in like that. I had a laptop, internet, a chair, and a spot in an actual hardened building. I felt like downrange royalty!!

I was able to do my rounds of going to each part of the data package each day, then 'hang my hat' in that office as my main spot. I would stay there until late at night, go to midnight chow, then walk home to my CHU. Then I would wake up the next day and do it all over again. And again, and again.

I made good friends with all of the contractors, and let me tell you, contractors are good people to know! Not only because they are smart and taught us many aspects about the network, they also didn't fall under any Army rules. They had freedom and they would drive all over the area and leave base when they wanted. These guys would go to Baghdad International Airport (BIAP) to eat at a Burger King. To the ones who couldn't leave base, this sounded like they were visiting another dimension. One day a contractor was about to leave for a BK run, so I gave him money and I asked him to bring me back a bag of the apple pies. I never had an apple pie like this before, but they came in a paper sleeve, and the crust was almost like it was deep fried. Inside had hot apples and apple cinnamon ooze. I was happy to receive the bag from him when he returned; I took the bag and visited all my Renegades to give each a pie. They were still super hot! Like apple ooze lava inside! Everyone loved them! I had one too and tasted amazing after only having Army chow for so many months.

One day we had personnel from our brigade in Kuwait visit us in Baghdad. They were coming to "check on the equipment." However, I had already explained how many people we had in charge of the one data box - the data box with many blinking lights. We had enough people there, each person could be in charge of one light. So, in theory we really didn't need them, but they probably just wanted to break up the monotony of life in Kuwait. Also, they could then say that they had been to Baghdad during the deployment instead of just staying in the safety of Kuwait all year.

In the group were two warrant officers, two staff sergeants, and one lieutenant colonel. I mean, warrant officers walk on grass, so they are definitely going to go to Baghdad if they want. They arrived and first visited the ITT building I was in, and not the trailer in the middle of the death field. Then to our surprise, one of my contractor friends offered to take us all to another base nearby to explore.

"Sure, why not?" I said.

It sounded like a totally safe idea. Drive out of the base, go through Baghdad to another base that I had no business to be at, and not tell my commander I was leaving... totally smart. But I was fearless, apparently, and I was intrigued. Let's do the tour! Joy riding just like in high school, except now the radio wasn't nineties hip-hop, it was Arabic chanting music, and we might get shot at or kidnapped. No big deal, let's roll.

My company commander would have lost his mind if he knew we left base. How did I know I was going to get back in again? I don't even remember how they checked us to enter the new base or how they cleared us to return. Maybe because we were in uniform? There was no ID scanning back then. However, it somehow worked!

We drove to an annex of Victory Base called Camp Slayer. It was at the southeast corner of Baghdad International Airport. There were many large man-made lakes on the base. It was unbelievable to see this and drive around a massive lake in the middle of Iraq.

We got out of the car to check out a long pier that had a rather large lake house shaped like an octagon at the end of it. I met someone who was walking on the pier; they told me they "lived" in the lake house. This guy saw that there was open room in the lake house, so he just brought his sleeping bag and belongings and started to camp out in there. It didn't seem so safe to me, with no lock on the bamboo door and screens instead of walls, but he seemed to like his new digs.

We drove to different spots around the base and got out of the car to take photos. My contractor friend took us to a spot to get a photo in front of the 'Victory Over America Palace' where, I was told, allied forces dropped a bomb with one hit on the leader of the Ba'ath Party, and it landed right on his bed while he was sleeping. You could see the remains of the massive explosion, and there were cranes from Saddam Hussein's previous construction all bent in the wreckage. They took a photo of me in front of the building, but I didn't smile big because it felt awkward to do so with a bombed-out building.

. . .

WE ALSO DROVE to the entrance of the huge building, with hopes of getting inside a little, but the entrance was filled with a lot of rubble and there were serious do not enter signs everywhere. We obeyed the signs but there was literally no one there so we could have walked inside if we wanted.

THEN ONE OF the strangest things I saw in Baghdad was the Flintstone House. Yes... Flintstone as in Fred and Wilma Flintstone. It looked like a massive life-sized papier-mâché art project of caveman dome houses connected by tunnels. My friend told me that they were made for Saddam Hussein's grandchildren and the area was used as a huge playground. At one point it even had running water and a working kitchen. Now there was graffiti all on the walls from the soldiers during the invasion only a few months before us being there - mostly names of the units that were there. (No, I didn't write "Renegades!" with a sharpie on the wall.) I took a few photos of this unique place, and I also took a photo from inside one of the little dome houses looking out the window at another house, from the perspective of a child playing.

Actually, this man-made hill allowed for a beautiful view of one of Saddam Hussein's son's palaces along the lake. While looking at the palace, my friend told me that the son was a Romeo / romantic type who always had a lot of women around him. Later on after the deployment, I learned that he was an absolute monster who raped, murdered, and tortured women in that palace. There was a balcony around the palace, which had many hearts cut out in the stone of the balcony walls as decoration. It was a strange contrast to look at the whimsical hearts on the party palace and a fun Flintstone children's playground, then seeing the surrounding rubble from the recent bombings.

We arrived back safely to our base, and no one even knew we had left. I had a taste of freedom, and I liked it. Still craving for some adventure, the next day, we decided to keep our tour local and went to the main palace right on Victory Base, called the Al-Faw Palace, and also known as the Water Palace.

The palace was huge, and it must have been the largest one in Iraq. While I was there I never referred to it, nor heard anyone refer to it, as the Al-Faw Palace. I found out later that Saddam Hussein commissioned its construction in the 1990s to commemorate the Iraqi forces' re-taking of the Al-Faw Peninsula during the Iran-Iraq conflict. I didn't know anything about that when I was there. I just chalked it up to be another strange and over-the-top thing built in Baghdad by Saddam Hussein.

Since we were with my contractor friend, it was no surprise we had the hook up to something not allowed normally! The palace was off limits. It was empty, gutted inside, and was being renovated for use again. The average person just couldn't just walk inside. However, we were with the ITT guy and a bunch of warrant officers, and I had my signal wig wags on my uniform, so it looked like we meant serious signal business. There were a few phone lines running inside and some antennas on the roof... so of course this super IT team needed to go check on them.

So, we walked in with no problem at all. The entrance opened to a large ballroom with big high ceilings and winding stairs going up to the next floors. There was a HUGE crystal chandelier hanging in the middle of the ballroom. As you walked up the winding staircase all of the crystals were missing from the chandelier, as far as an arm could reach and grab them. We couldn't even take one, even if we wanted to, because the soldiers from the invasion before us took all of the crystals that they could reach.

Other items that you could tell were there at one point but taken, were the much talked about golden toilet seats. I saw pictures of people from the invasion sitting on the toilet with the golden toilet seat as a joke. Then I guess when they left they must have taken them home. Lord knows where they are now. Maybe they will show up on eBay one day. Maybe they will show up in a military museum at one point. Or perhaps they are already there, and I just haven't heard about it.

I was able to take an epic photo on the famous palace throne. There really was a throne in the ballroom area, and I can only assume it was left there for people to take photos in it. Everyone called it Saddam's throne, but I don't know how true that is. Let's just go with it. The throne had a high decorated back and cushions on the headrest.

We followed the contractor up the stairs, passing many empty offices and rooms completely gutted with no furniture whatsoever. We went to the top floor and then there was an opening to the roof, a slim shaft with a long ladder to the top. This was the ONLY way up and down from the palace roof. It was a pretty high ladder too. And pretty high for me, someone who never climbs even small ladders.

Did we want to go up this one entrance/exit ladder to the top of Saddam's palace and hopefully not get shot by a sniper, to get a good photo? Of course, we did!

Someone actually took a pretty cool photo of me climbing up the ladder. If I did fall to my death on the way down or get shot by a sniper on the roof... at least there would be that as the last cool photo of myself.

I PERSONALLY SAW several items that were looted from that palace. A warrant officer preparing for his return trip home showed me a large, gold-framed painted portrait of Saddam Hussein. He had someone take a photo of him in front of the palace holding the portrait so they would know he really got it in Iraq and it was legit. I also thought this was self-incriminating because we aren't allowed to take war souvenirs with us! Like I said before... warrant officers walk on grass.

Another person showed me some nice, hand-painted china they took from the palace. I guess they were planning on having a nice dictator-themed tea party when they got back home from the deployment.

We gave the palace back to the Iraqis in 2011. Sometimes I wonder what it looks like today... and I will keep wondering (or I will have to Google it) because there is no way I will go and see for myself in person.

Contractor Shane came in to help so many times. Once my battalion commander came to visit us in Baghdad. She was an amazing leader and officer. We knew she was arriving at the airport, and then would proceed to Victory Base with the convoy team. Unfortunately, when she arrived, the convoy had already stopped picking up people for the night and she was stuck at BIAP in the extremely beat up and flimsy temporary tents at the airport. They were also dangerous because of they lacked protection from explosions.

We got the call that she was there while my company commander was visiting me in my bougie office space. Fully knowing she was there already waiting, he said, "Well there are no more convoys, so she can come tomorrow."

"Come tomorrow? Why don't we just go pick her up?" I asked from my executive desk.

He didn't know that my confidence and adventure spirit was high because of my joy-riding escapades, but that's beside the point.

He looked at me and left to go to bed.

It was time to take matters in my own hands. I turned around to look at Shane, and I didn't even have to say anything to him, we just made eye contact. His desk was behind me, so he overheard the whole conversation that just went down.

He looked at me, grabbed his Kevlar helmet and said, "I'll go pick her up."

Shane to the rescue! Shane is the man. Really... unbelievable.

I waited in the office, he brought her to our area, and we got her temporary lodging all taken care of. She looked extremely tired and also very relieved that she didn't have to stay in the temporary tent at the airport. You literally get no sleep in those loud temp tents.

She was a classy and professional officer. She was one of the first females to go to West Point from the state of Montana. She always looked presentable and nice. She had long blonde hair, but she always had her hair up in a French braid.

I truly was grateful for Shane's assistance that night. The icing on the cake of me being able to mingle with the contractors in this building was the luxury of having a bathroom in the building where they let me take showers. It was a big room, but like the majority of the building, the bathroom was not finished. Not only were tiles missing, but we could tell other things were missing that they had intended to add to the room. Perhaps there had been plans to make a very luxurious, spa-type bathroom.

In comparison to the shower trailers, this felt like another world. The privacy and solitude that I had in this room was an incredible feeling after not having it for so long.

Another visitor came, but he wasn't coming to just visit us. My new platoon sergeant, Sergeant First Class E, came from Kuwait to stay with us in Baghdad for the rest of the deployment. The first sergeant also came along with him. They flew in because, at this point, they stopped conducting the long convoys, like the one we did to get to Baghdad; long convoys were too dangerous now. If you had equipment to bring with you, it was line-hauled by a contracted truck company.

I was very happy to have Sergeant First Class E with me and all the other Renegades in Baghdad! I felt like we were more of a complete team with him there. He always had my back, and I felt that we could now accomplish much more together.

BAGHDADDY RESIDENT

There were literally shirts for sale in the Kuwait PX with "Who's your Baghdaddy" printed on them, and I heard that Army wives back home complained about them because it was hinting to a sexual reference. Also, insinuating who is your Baghdaddy only in Baghdad...dy. It is actually a ridiculous thing to sell in the PX, but back then it was very funny. Gosh, now I wish I bought one.

We didn't have a PX on Victory Base to get epic shirts like that, unfortunately. We did have a small shop that was in a former guest house. AAFES got a bunch of shelves and set them up in this small apartment with the basic essentials items to purchase. In one of the small rooms they had some local craft items like beautifully carved wooden decorations and metal lamps.

We had a lot of DVD buying options, but they weren't super cheap and storage space was a problem; therefore, you had to be selective with your choices. There was also a lot of snack food like chips, dip, candy, and beef jerky. There were also some pretty Iraqi boxes with mother of pearl designs on them. I purchased a few and still have them to this day.

Once I went to the BIAP PX on one of my secret joy ride operations, and I got a single DVD player. I thought I was so cool after getting this because I could watch a movie all by myself in my room without a big TV and DVD player! This was the peak of technology at that point. It was the same cool, cutting-edge feeling like in college when I had a portable CD player (hold it flat!), but this was a freaking DVD player!

THE BOOTLEG DVD

Another new aspect of technology was buying bootleg DVDs with multiple movies on them. Either they were movies that were already out, and there were three to five movies on it with the same movie theme, or it was a new movie still in the theaters and it was a bootleg version. You would see the top of people's heads in the theater and see them stand up to go buy popcorn. Most of the time the quality was terrible, but somehow we didn't care.

These movies weren't sold officially by AAFES, they were sold by the knock-off DVD vendors on or around base. Basically, everyone had some of these movies, and when you were done with them, you could swap them with someone else.

My soldiers were complete angels of course, but the unit from Fort Hood was constantly getting into trouble. Big trouble, not little trouble. The lieutenants from the unit would tell me all the info. Someone got caught selling drugs; I didn't understand how that was possible because we couldn't even get blue Gatorade there, but, somehow, they were getting drugs. Impressive.

Also, I heard there was a female officer who was PIMPING out soldiers in the CHUs. I mean, where do you find the time and energy? CID got involved and busted the whole prostitution ring. How boring am I though? I was watching 'Girl with a Pearl Earring' on my single DVD player for the twentieth time and this lieutenant was out there pimping as a side hustle.

Another female behaving badly, and not too knowledgeable about the rising DVD technology, also got in trouble. She made a porno-type movie that some guy recorded. She most likely got paid for making it. Then the guy who recorded it sold it to those knock-off DVD guys, who then made mass copies and sold it on the base. Then the brigade commander heard about it, called her in, and played the DVD while she was in there so she could confirm that it was actually her. If this is all true, how embarrassing and mortifying for the soldier! I'm not saying that it was a smart decision to make a video like that, but she most definitely didn't think it was going to mass produced.

Another instance of someone getting in trouble was a soldier and a contractor when they got drunk together. They stole a Humvee and drove around the base while they shot off some rounds from the soldier's M16! Their night of joyriding came to an abrupt stop when they crashed the Humvee. I didn't see it because they did it in the middle of the night, but when I heard about it the next morning, I was totally shocked about what had happened while I was snoozing. There's drunk, then there's that level of drunk where you steal a Humvee and do drive-by's in a war zone. I was just grateful that nobody got hurt.

OFFICE SPACE

If I haven't stated it enough, I really enjoyed the area I worked in because I felt safe, I was around great people, I had a laptop I could use, and there was always the hustle and bustle of people coming in the building, which broke up the routine.

Mail was delivered right to my desk because one of my soldiers would go on mail runs in his Humvee to the back of the base and bring our mail to us. He enjoyed this because it was something to do, and it was an exciting trip. There was a large tent on the edge of the base and in the tent there were bins marked for each unit where letters and packages were sorted. Hell, yeah it was exciting to get mail there! Some days the soldier couldn't go check mail because the whole area would be blocked off because of a lot of mortar fire coming down and it was too dangerous.

Amazon wasn't a thing yet, so we didn't really order items online. The only packages we would typically get would be care packages from family and friends. Sometimes you might get a care package addressed to the unit you were with, but not too often in Baghdad. Once I received a big box of Girl Scout cookies from the troop back in Germany. That was pretty great to receive. I took a photo with some people from the ITT building and some of my soldiers to send back to them as a thank you. A lot of the daily hustle and bustle was centered on people getting packages or sharing care package items with everyone.

Along with all the goods trading, our office was a rotating door with people coming in all day long. It was pure entertainment at times. We didn't have a TV and honestly we didn't need one. Since my commander was way out in the middle of the field, people came to talk to me first since I was right there in the mix of it. It's not that I can't remember this woman's role now, I didn't know it back then either... but some major came in once ranting to me and everyone in there. She was ranting and yelling loudly about how important communication was and that if the comms went down then, "People will die!!" She had a point, and I appreciate her appreciating the Signal Corps, but I believe she was in the middle of some mental breakdown.

Another incident I won't forget was when I received a phone call from someone who wanted to come talk to me about his unit's communications. He called, telling me he was Captain So-and-So. I was a first lieutenant, so we are basically bros. I was pretty relaxed and told him to come on over so we could talk about it.

"Yeah.. you come to me. Make it snappy cause I am going to my meal number two out of four soon," I said so cool to my fellow comrade-in-arms.

That was the day that I learned that a Navy captain is not the same as an Army captain.

(In my defense, homie never said he was Navy.)

Shortly after our casual conversation, in walks an older man, with the same rank as a FULL bird colonel.

I could have died.

As a first lieutenant, I would never have a meet-up or conversation with a colonel. This would be like me chit-chatting with my brigade commander... pretty rare to never. I went from "too cool for school" on the phone to "Military Bearing G.I. Jane" in a matter of a few of his footsteps in the room. I stood there talking about whatever his question was, and all I could keep thinking was, "Holy shit, I just summoned this Colonel to my office."

LIVING IN DANGER

When you are in a combat zone you receive combat pay or hazard duty pay, which is a couple hundred extra a month. For a young lieutenant or soldier right out of OBC or AIT, having an extra couple hundred dollars is a big deal. We received this pay for being in Iraq and also Kuwait. Even though the locations were very different in the level of being in harm's way, they both fell under the combat pay entitlement.

Incoming!

What separates regular Army life from a combat zone? Well in a combat zone, people are trying to kill you - at all hours of the day in different ways. You are an invader to them in their country, and they want to kill you and as many of you as possible. There are no time-outs or safe zones on the base. When you go to sleep at night, you might have a mortar land on your CHU, because the enemy knows where you sleep. When you go to eat, you also might be attacked because they know that is where we all go to eat. Walking around the base, especially on the perimeter or on the roof of a building, you might get shot by a sniper. Getting close to the main gate, all we could hear were gunshots over and over again; we also heard this coming from the towers. While it was not as frequent, every so often a rocket would find its way to the inside of the base.

I was sitting at my computer in the tech building one day, and I heard the loudest sound I ever heard in my life. Not only did I hear the sound, I also FELT the sound. The vibration went right through my chest and entire body. A rocket hit right behind our building! They said most likely the enemy was aiming toward the big tower that had a blinking red light on top, but missed slightly.

Everyone put on their full battle rattle and ran out to go see what happened. We met the man who parked his SUV right by a port-a-john to use it, and when he was in there, the rocket hit. He was safe because of the Jersey barriers around it. This man gave me photos (via USB swap) of his bombed SUV. I received photos of what his SUV looked like when he came out of the port-a-john.

The Hole of Almost Death

We often operated in black-out mode, where all lights were turned off so that the people mortaring us at night would stop aiming at our sleep trailers. There was a 10-foot deep cable vault right outside our technical control facility (TCF). From the outside, it looked like a huge manhole when the cover was closed. Picture an opening almost three or four times larger than normal manholes. There was a small ladder on the side of the wall to climb down into it. The bottom was all concrete and planks were set up because there was some dirty stagnant water in there from the ground water.

One dark night a young airman was working in the cable vault trying to fix some broken wires. Not knowing that he was out there, I walked out of the TCF on the way to midnight chow (because why not?), and since this was right outside the door in the path to get to the road, I walked right toward it, like I always did. I was walking with a battle buddy, of course, but they were a few steps behind me. I stepped and had the strangest sensation where I could kind of feel that there is nothing under my foot. I was able to sense the vastness of the floor missing. I stepped but THANK GOD I didn't fully put all my weight down and I was able to rock back real fast.

The airman must have decided to go to midnight chow, but he left the cover to the vault open and didn't put out any orange cones or danger tape around it to block it off. There was no warning at all, and it was impossible to see in the darkness. I was in shock afterward thinking what could have happened to me if I fell in that damned concrete pit. And honestly it would sound so stupid, and people would have made fun of me. I would have been the lieutenant who wasn't looking and fell into a hole. Lieutenants had a bad reputation for getting lost and doing stupid things, so falling in a hole was not going to help me at all.

To this day when I see subway or sewer grates ... the kind where you can see how far down they go... I still think about how close I came to falling down that far onto hard concrete. What the hell would have happened to me? Would I have landed on my feet and then my hands? My side? My head?!

SOLITUDE IN THE PITCH-BLACK NIGHT

Walking back at night really scared me because it was pitch black and also very quiet. Since we stayed up so late at night to avoid the mortars and boredom in the trailers, I walked back around 0200 most of the time. There was always talk of people coming over the walls and breeching the perimeter. Also, there were contractors and locals out there sexually assaulting women. I was on high alert, my heart would basically race until I was in my room and had locked the door behind me. Then I would only have to worry about a mortar or rocket landing on my locked room. No big deal.

Walking home one night it was especially dark and quiet for some reason. I walked along dirt paths with palm trees, walkways with abandoned buildings, and then past rows and rows of trailers and bunkers in the sand. Perhaps we were on high alert for people trying to come into our perimeter, but that night walking on the road alone I was so frightened I held my magazine ammo in my hands and close to my chest if I needed it. As I was walking on the hardball road, a civilian car came driving down the road slowly behind me. There were very few cars then! No units at this point had cars, and those who did have cars didn't drive on the base with them. It was basically set up as a walking base then with more paths than roads... not for cars.

The car drove slowly behind me, and I was walking but feeling petrified for some reason. It was 0200 and here was this twenty-four-year-old blonde girl walking by herself on a road on a base in the middle of Baghdad and holding a ridiculously long (so therefore useless at close contact) rifle. The car passed me, and of course I had to look over my shoulder as it approached. I must have looked scared. In the passenger seat was a pretty blonde woman, and she smiled warmly at me, as if she was saying, "Don't be scared."

Then they were gone. I felt kind of silly for feeling so scared of the car. Then I thought, "Who the hell was that anyway?" We didn't have civilian women there with their hair down. It was a surreal experience, and it felt like she was a guardian angel smiling and nodding to me that I was going to be okay. Maybe she was? Or maybe she was a news reporter? I will never know...but she did give me a few moments of feeling calm and safe.

BESIDES ALL OF the other dangers of war around us, we also had to be cautious of our own people on the base. Not the enemy on the outside trying to come in the perimeter or sending things in the air to land and kill you, but also Americans walking in the dark on your own base.

When I was there, a female soldier was attacked at night by a contractor who drove the supply trucks and stayed on the base. He attacked her from behind, grabbed her slung weapon, and pulled her back. He then pushed her into her own room and sexually assaulted her. So, this massive M16 that we have slung to our backs, with ammo in the cargo pocket bouncing around and hitting our calves, is not only pretty useless (unless they are going to pop up in a distance like targets on a shooting range) it is also adding to our vulnerability because some horrible person could tug hard on it and drag us to the ground. Just great.

The day we found out about the attack, the other female in my platoon and I both bought switchblade knives to have them on our belts. If some evil contractor tried to grab us, he was at least going to get stabbed.

My platoon sergeant was immediately shocked by the news of the sexual assault, and from that day on he walked back with me to my trailer every night to make sure I was safe. That meant so much to me; looking back, it still means so much to me.

When Sleeping is Dangerous

The time spent in my CHU was never enjoyable. I never slept through the night well. I would stay up late in the OPS area with my platoon sergeant and then go back to my room around 0300 or later because there was less of a chance of mortars. I wasn't sleeping anyway. Nights were when the attacks were getting so bad. Also, I was completely and utterly exhausted. I got to the point of being so tired, I just didn't care. I pushed myself to the point where I had to go to sleep. My brain and body were shutting down.

One particular night I came back at 0400. Early the next morning I was awakened by the sound of gunfire and then mortars falling in the distance. Those bastards AIMED for our sleeping areas. And guess what? I just didn't care anymore.

I didn't want to die, but I was so broken down tired from not sleeping for so long, that there was no choice but to sleep. I put my Kevlar helmet on my head and put my Kevlar body armor on my chest. I curled up into the body armor so that all of my internal organs would at least be covered a little if my CHU was hit with a mortar. I felt so sad that it has come to this. I was so severely exhausted that I was ready to just hope for the best if the mortar hit my CHU. This was perhaps the closest I have come in my life to accepting the possibility of death.

Times were getting worse and more dangerous by the day. I did things completely out of character. I didn't eat snacks back then at all because I was really concerned about my weight, and it was just not me. However, I got a bag of Sun Chips from the small PX and ate the whole bag. Like I said, this might be normal for some people, but I never did this. I guess I felt that it didn't really matter because I might die one of these nights like other people, so I might as well enjoy this bag of French Onion Sun Chips.

Something else that I can't explain, but that is still with me today, is my love of the color purple. I would half-sleep at night and feel scared and worried. When I closed my eyes, I would see beautiful dancing, deep amethyst purple colors. They were swirling and moving like a beautiful purple aurora borealis in the sky. Was this how my mind felt it needed to protect me? Or maybe it was something not from this world that I can't explain. But it was soothing and calming for me, and it gave me a sense of relief. Since then, I have been completely drawn to purple and have collected many items in the color purple.

Another symbol of the sad aspect of living in danger, or a combat zone, was the index card that I put in my top pocket in my uniform. It was my "if I die card." I had a one-liner good-bye for key people in my life. I also had my America Online (AOL) email password and all the passwords to my banking information. I did this to make it easier for everyone to get in my accounts if I died. Looking back, I don't know why anyone needed my AOL email though...maybe to forward some good luck chain-letters one last time for me.

Luckily no one ever had to find this note. Others weren't so fortunate.

When it comes down to it, most of the time it was a complete random crapshoot of being in the wrong place at the wrong time. Many of the people who died when we were there could have been anyone from my platoon. In other cases, their jobs and duties were in higher risk areas, for example being on convoys.

HORRORS AROUND US

I was witness to some soldiers being severely injured during my time in Baghdad. This is something that never leaves you, and you do not forget. People that I know in my life now would never guess that I saw things like this during my time in the Army. It is something I don't talk about, but I do think about the well-being of the soldiers and also the families who were affected. I don't even know if they lived.

I was leaving my CHU one day, walking along a path to where my soldiers were located. Then all of a sudden, a flatbed truck came storming up the road and stopped right where I was standing. I was set back a little, and there was a sandy area in between us, so I wasn't directly standing with the soldiers. The flatbed was driving so fast and erratic, and there were three soldiers on the back of the flatbed... something was not right at all. Two soldiers were sitting upright, and one was laying down. This isn't how soldiers travel on vehicles in the Army. The flatbed came to a quick halt, and then I saw the outline of the soldier who was laying down: he only had one foot up in the air. His other leg was missing.

People from the small hospital came out running and took him on a stretcher. His two battle buddies were left there standing alone, and their faces looked overwhelmingly devastated; they were terrified and in shock. I felt what they were feeling. What did they just come from and what did they just see? The feeling in the air was like the aftermath of a car accident when everyone is in an adrenaline shock.

I couldn't walk anymore. I was frozen and overwhelmed by what just happened right in front of my eyes. I had never seen anything like this before in my life. I felt a sense of guilt that we had it "easier" on the forward operating base (FOB), and other soldiers were out in the perimeter getting hit with IEDs and shot at.

Another time I was in a small building right outside the medical building that had a telephone in it. A woman was in there screaming and legit freaking out calling someone. She was a civilian, and I don't know what her role was there. She was yelling in the phone that they just got attacked and a soldier was killed. She said that they were driving in a tactical vehicle and a civilian car passed them, then someone popped out of the sunroof and sprayed their vehicle with bullets. She said the soldier was shot and bleeding everywhere and dead right now in the back of the military vehicle. All of this was said totally frantic and very loud.

So, I felt like I shouldn't be in there overhearing this anymore and I left the building. Then I walked outside and right in front of the door, parked haphazardly, was the vehicle she was talking about. I was dumbfounded that the vehicle was right there... and the bleeding dead soldier was, too. I remember feeling so sad for him, and also sad that he was alone in the vehicle. I knew he was dead but somehow I felt that he should have been watched and guarded in a way. He was so far away from his country, and this was how it ended for him. Also, his family back home didn't know anything yet. But I knew. The frantic lady knew. And whoever she was reporting it to also knew. That's all.

There was another instance where I had to go into the medical building, and since it was literally right next to my CHU, it wasn't rare to be there. Another one of my soldiers was sick, and I went back to the area with the small, curtained rooms. When I was walking out to leave, the main room was suddenly turned into a triage area because they pushed in a wheeled table with a man on top and started treating him. About five people surrounded him, and I was right there on my way out. His shirt was off, and he had bullet holes all over his body. I could only see his upper torso, and I only looked quickly by natural instinct. I did not mean to stare or look; I was just reacting to the chaos that just entered the room.

I had never seen anything like this in my life. I know the military medical personnel are used to this, but I had never seen someone shot before. It looked like a scene from the movies, with black dots all over him dripping with blood. I stopped in my tracks. I know I shouldn't have but I explained why I did. Someone looked at me and shooed me off with their hands. My trance was broken; I got the hell out of there. Did he live? I will never know.

I did talk to a soldier who got shot and lived to tell the tale. I was getting my haircut, which was a crazy experience in itself. You may be asking yourself why the hell would I be getting my hair cut in a combat zone?! My hair was getting long, and my bun was getting very heavy. Unlike today's rules, we couldn't have our hair down in a ponytail and it had to be wrapped up in a bun on the back of the head. It kept falling out, and I felt the pull on my head. So, I just needed a trim, and on Victory Base there were no salons for women, of course, so I went to the men's trailer where they got buzz cuts. The man cutting hair was from India, and he seemed very timid and cautious cutting my hair. I needed just a straight cut in the back so that it was shorter... a few inches off. It was no doubt the strangest haircut I ever received. He softly took a small amount of hair between two fingers and trimmed it very lightly. He repeated this very slowly from shoulder to shoulder. In hindsight I should have just cut it myself with some SKILCRAFT office scissors.

Anyway, the trip there allowed me to meet some interesting characters! There was a crew of infantry guys in there and a command sergeant major was showing off his soldiers. He introduced me to his "Bad Ass NCO" who got shot through the neck. Not IN the neck, THROUGH the neck. He was on one of the watch towers on Victory Base and he was shot at from outside the perimeter in Baghdad. Somehow the bullet didn't hit anything vital, and they were able to stich him up. Then he wanted to come right back here with his team! The command sergeant major was damn proud. I saw the scar... well both of them.

THE SHADY SHOP

Way in the back, along the perimeter, of Victory Base was a shop that was not much more than a big shack with a tin roof. There was all sorts of junk and cars around the sides of the shop. Honestly, the best thing we could call this was the Shady Shop. The owner was a friendly, yet shady, man from Lebanon who always seemed to be making some business deal; we felt like he was never telling the full story about anything. He had a few other guys working with him; one was a nice young Iraqi man named Mohamed. There was another guy working there and I don't remember his name, but he was Iraqi who had been living in America for a while before this. All three lived in downtown Baghdad, which kind of blew my mind, because for me it was the center of enemy territory.

But they went there at the end of the day to their apartments and slept.

"We just don't go out at night, and we just hope for the best that a bomb doesn't go off when you are out in the city," Mohamed informed me.

"What a way to live life," I thought to myself. Even though that was kind of how we were living on base; at least this was a temporary situation for us.

I can't deny that the Shady Shop owner liked me extra because I was a friendly female soldier. He would give me drinks like juice that had little chunks of mango in it and tasted so good. I never had anything like it before. But thinking about it, I wondered if I really should be taking drinks from the Shady Shop owner in the middle of nowhere along the perimeter of an Army Base in Baghdad. He could have drugged me and sold me on the Baghdad black market. Luckily, he didn't.

This guy could have ordered anything you wanted. It was like the saying, "I know a guy." Well, this is the guy. Once I was in there and a colonel was working out some deal for a bunch of SUVs, and he gave him a credit card. SUVs. A bunch of them. Just charge that shit... no problem. Oh no, that's not too shady...

You wanted some refrigerators? No problem. How many dozen do you need?

He was clearly making bank. The other man working there who said he was half American or something to that effect, told me that he stayed in the same apartment as the owner, and one night the owner ordered six (six!) prostitutes to his place. When he came home, he walked into the party. Then he told me that he just went to bed in his room.

"Riiiiight... suuure dude," I thought to myself. I guess he

thought I was pretty gullible to believe that. Anyway, let's go back to that... six! He was making so much money on selling bulk swivel desk chairs, he decided to do a Costco bulk order of prostitutes, too!

Mohamad was a very nice and kind person. He told me that his father and his brother both died in the war against Iran in the early eighties. I felt embarrassed that I didn't know about this previous war while Mohamad was living without his family members forever because of it. I couldn't imagine what that must have been like for him. We came from two very different worlds for sure. And here we were on some middle ground on the Army base becoming friends. He was one of the only Iraqis that I ever came in contact with and ever talked to.

Mohamad hooked us up by getting us the new hot topic around the base: the new Iraqi dinar! The word going around was that the newly printed Iraqi dinar was going to be like the Kuwaiti dinar was when that was first introduced. It was sold for a very low price, and later the exchange rate was almost one to one with the dollar. So, if you bought a million Kuwaiti dinar and it went one to one... well you can get do the math. We were more than ready to become millionaires.

"Mohamad, can you please get us some Iraqi dinar? It is all the talk around base, that in a few years we can be millionaires!" I asked excitedly.

"Yes, no problem, you bring me the dollar cash, and the next day I will bring you the dinars from the city," he happily replied.

Most of us bought a million dinars which cost about $600. Think about the trust that I had with Mohamad, and he was not making anything in return for it. I continuously gave him $600 dollars to take to exchange for my soldiers, when he easily could have denied I gave it to him, or just disappeared with it.

We didn't want to get in trouble with customs or any officials, so some people put it inside the small storage compartment in buttstock of their M16 rifles. The fact that Mohamad got it for us from the bank, we got the best deal possible. Unfortunately, the dinar didn't skyrocket, but here's to hoping that one day it will!

I didn't buy much from the Shady Shop, but one thing I did buy was a whole bunch of the money before Saddam Hussein took over. It was beautiful money, with pretty artwork on it with horses and other animals. Then I also bought some money with Saddam Hussein's face all over it, which was no longer valid. Mohamad told me that when Saddam Hussein took over, he had all the money changed and had his face on every bill.

Visiting the Shady Shop started to become very dangerous, so the mango juice cocktail parties had to come to an end. The mortars started coming in more and more, especially toward the back area because it was so close to the edge of the base and in the direction of the city.

I remember the last happy hour there, standing inside the shop, talking to my friends who were on the other side of his long, glass counter. Items for sale were stacked to the ceilings and along every inch of the walls. Out of the corner of my eye, I saw his taped up messy collage with the smiling faces of his wife, kids, and grandkids. An image of his six-prostitute party came into my mind, which I quickly shook away with rapid eye-blink and head shake.

The mortar attacks prevented me from saying goodbye to my friends there and thanking Mohamad for his friendship and kindness. Our area to roam became much smaller, and we were back to using what we had in our area or what we received in care packages.

BAGHDAD BARISTA

My family sent me some fancy Dunkin Donut coffee in a care package. This was a big deal because we didn't have coffee shops anywhere. I wasn't really a coffee drinker back then, but the only place you could get coffee was in the dining facility, which was extremely basic. The other choice was to make it yourself if you brought a coffee maker. So, when I brought in a couple of bags of Dunkin Donut hazelnut coffee beans, it was pretty damn exciting. The building where I had my desk had a coffee maker in the back room. There was a coffee bean grinder, too, so I felt like the real deal grinding those beans. I wanted to surprise and treat all the IT guys in the building, so I went back there and brewed up some coffee. I didn't know a damn thing about making coffee, so I read the ratio information on the package. "Okay, got it. This many scoops of freshly ground beans. Okay now add the water. Too easy."

A few minutes later, we had some freshly brewed Dunkin hazelnut coffee wafting in the air. The results were what I expected. Imagine a cartoon where the characters smell something and then float into the room. Everyone came in the back and said, "Where did you get Dunkin coffee?" "Yes, I know... you. are. welcome." *pats my own back*

In the spirit of good comradery that night, I also drank some coffee. It was a wonderful moment of bursting flavor and a taste of home. I drank it slowly and held it with both hands wrapped around the mug like I was in a Folgers coffee commercial, taking the first morning sip. We all did.

When I walked back to my room that night, I felt like someone punched me in the stomach and I had pains like I never experienced in my life. That part wasn't in any Folger's commercial.

It turns out that there was some information missing.

Some very important information that I didn't know about at the time of brewing.

The next day, someone posted warning signs all over the buildings, on the sinks, and on any water source: "Do NOT drink the water."

Oh no.

I then realized that, not having this tidbit of information, I had served up some hot brew of polluted Iraqi sewer, knife-stomach-juice to a building of hardworking nightshift workers. I can only guess everyone else had a similar night as I did. If you are reading this now...

...Damn, I am sorry.

MY SECRET RIDE

I made friends with people very fast. I was a young, friendly, funny, female from New Jersey. I mean, what is not to like really? I made friends with all the contractors there quickly. They were kind of assigned to my platoon to fix the stuff that we weren't able to. Some of the equipment even came with a handful of contractors, and their job was to only fix the equipment if it broke; we weren't even allowed to touch it. They were smart, and they got paid A LOT of money to be there. They were also not under any sorts of rules. One of the lead guys really got along with me. He gave me the keys to my literal freedom. He gave me a Gator, which is a four-wheeler, and this one had no top. A convertible badass Gator if you will!

Here is the zinger: My commander didn't know. I knew that if he did know, he would tell me to give it back. My platoon sergeant also told me that this is what would happen if the commander found out. So, it was now a secret ride. I had it parked in an obscure place, so I didn't have it parked outside my living area. I didn't have twigs and branches to cover it like in old James Bond movies, but it was pretty close.

Two people could ride on it, and that is what we did. I drove around with my female soldier a bunch, and also I let my platoon sergeant ride it as well. The name of the game was to not be seen by the commander. I am not sure if he ever found out about it, but he for sure never saw me rolling by.

The Puerto Rican National Guard was stationed there, and there were a lot of them. Toward the edge of the base, they had an empty building they used for morale, welfare, and recreation events. Sometimes we couldn't go there because a sniper would have been spotted nearby and would take shots in that area. Then after a little while I guess they would forget, and it would be allowed again. So, the PR National Guard hosted a... dance night. They made flyers, the word got around, so we went. We took the secret Gator to the back of the sniper zone in Baghdad to learn to dance the Salsa from a massive crew of Puerto Ricans. Yes, that sentence sounds crazy to me, too... but it happened.

We rolled up like bad asses on the Gator ready to one, two, cha, cha, cha. I did learn some Salsa, the Bachata, and Merengue. I loved it! They did it a couple of times during their time there. If I go to a Salsa club to dance now and someone asks where I learned how to dance, I have a great story to tell them.

The dance club area also attracted a lot of hip-hop dancers and young Black soldiers. They played some rap/dance music, and I never saw people dance with each other like that before. And it is not like I grew up in some backwoods town or anything... but the girls were grinding and gyrating on the guys, and the guys were basically only humping them. (What dance lesson session is this on the flyer?) They would do this in a circle of dancers. I had that feeling that I should not be there, and this was awkward AF. Also, everyone had weapons on their backs by the way, to add to the strange situation. Then at one point a first sergeant stopped everyone from dancing so close and humping, and just like that the awkward vibe was gone, and everyone went back to the Latin dancing.

Also, some Bangladesh workers were hanging out and watching. These guys would hang on each other, hold hands, and pet each other just to get a rise out of the US soldiers. The soldiers would laugh like crazy or point and say... ewhhhhh! However, the Bangladesh guys thought it was so funny that the US soldiers didn't like when they would pet and kiss each other, so they just kept doing it more for the laugh. This side situation just added to this strange planet / club I just arrived at in the middle of Baghdad.

International Woman of Mystery

Since I was stationed in Germany and had been to a few USO trips to other countries, I considered myself as an International

Woman of Mystery! Not really, but I did feel worldly for sure and when I saw international military members walking around the base, I felt compelled to be the welcome wagon and bridge the nations together.

This base was formed to be a combined joint task force (CJTF) so there were many units from other nations. There weren't that many different countries represented, but they still stuck out a lot. They all had their own trailers for their headquarters, and they were all in the same area we called International City. Each trailer had the country flags waving on top of them. I actually never went to the International City, but it was on purpose. I don't think any US soldiers would have walked through their area to just look around. It definitely did not have that welcoming feeling.

So, one sunny day we were walking back from lunch and were passing a group of about five to seven male soldiers from Spain. How cool is that? They looked so nice in their uniforms that I wasn't used to seeing. And since I was so confident in my high school Spanish, I decided to give it a whirl. I am about to impress my fellow soldiers I am walking with, and I am going to make some new Spanish friends. I am going to say hello in Spanish.

So, here I go, "Buenos Dias!"

Insert a little wave and a big smile.

Then I got the most unexpected reaction from one of the taller men with long, black, curly hair. The dude practically jumped out of the little pack they were walking in and, with a very annoyed and tense voice he said "Después de almorzar, no es buenos días ... ¡Ahora es buenas tardes!"

Which means, "After you eat lunch, it isn't good morning ... it is now good afternoon!"

Here you don't insert a little wave and big smile. He was clearly pissed at me.

You can insert the motion of eating a sandwich. When he was schooling me on my Spanish manners about lunch time, he did the motion of him eating a little invisible sandwich.

I was shocked. I was actually pretty freaking embarrassed. And you bet your sweet sandwich nibbling ass that I didn't say hello to him/them again. God forbid the whole, you know, "combat" and "war around us" makes me forget what freaking time it is.

Little did I know that I would get assaulted by the Spanish grammar police while walking back from lunch.

THE ITALIAN STALLION

One fine Baghdad day I made friends with an Italian soldier named Fabrizio. He was a technical warrant officer who used to be a K9 dog handler. I thought it was awesome that he was from Italy, and he was just really down to earth and funny. He was sort of on his own there and didn't have a large unit that he was attached to. So, we ended up going to meals together. I would also bring some of my soldiers, and we all became friends. He didn't really speak English so much, and I only knew a little Spanish and no Italian, but somehow, we got by. He actually learned English SO fast just by hanging out with us. It was really impressive.

We went to midnight chow together a lot. Midnight chow is exactly what it sounds like... it was a meal at the dining facility at midnight. The main purpose of this meal was for everyone on night shift. Honestly, everyone went to it out of boredom. We would stay up late at the OPS area, maybe surf online, then go to midnight chow, and then to the rooms to put on a DVD to fall asleep to.

A movie saved my platoon sergeant's life one night. One night he said he wanted to go to his room early. It wasn't usual, but I said good bye and good night. During his walk back, he stopped at the small PX shop we had on the base to buy a new DVD. Then he continued on his walk to his CHU. When he arrived, he found his CHU completely blown out because a mortar just dropped right on it! The mortar actually dropped on his bed and then exploded shards of metal through everything in his room. He only missed it because he went to get a new movie. If he didn't stop, he would have been sitting on that bed. When he arrived, everyone was looking around for him in the smoke and wondering where he was. Then he walked out from the shadows looking as shocked as they were... holding "Crouching Tiger Hidden Dragon" in his hand.

So, that's another reason we went to midnight chow. Less time in the CHUs.

The food was really good, too! The workers were from India, and it seemed similar to a catering company. This was very different from other Army dining facilities because normally they are run by the military with bulk Army food.

I would have these awesome grilled cheese sandwiches at midnight chow, but the options were endless, and it was a judgement-free zone. Also, they had an ice cream bar with a server there ready to scoop for you. Again, nothing like a regular Army dining facility. They had mango ice cream, and it was so delicious. I mean, after having bland food in Kuwait and meals ready to eat (MREs), mango ice cream was out of this world for my taste buds.

So, Fabrizio was my new midnight chow buddy. He would tell me what it is like living in Italy, and I would tell him about dirty Jersey.

"I live on a large farm in Tuscany, with many pure-bred Italian horses that live on the property. We have grapes vines for making our wine and olive trees to eat fresh olives with the wine," he explained to me.

"Oh, that sounds pretty amazing... last house I lived in a New Jersey was a Rutgers fraternity house, and we had an opossum that lived in the basement that would scare us when we tried to do laundry. Then down the road there was this parking lot with these trailers called 'Gut Trucks' which made greasy French fry-cheese steak-mozzarella stick filled hoagies," I replied with a strange comparison.

Somehow his sounded nicer.

He would come to the building I worked in, and we walked to the dining facility together. It was creepy walking around there at night, and especially to the dining facility. Along the walkway to the entrance there were a lot of small cages on the grass. We found out that Saddam was making his own zoo to impress people who would be walking to the building, which was now our dining facility. I heard that there were animals like lions, tigers, and monkeys. Also, when the invasion happened, everyone abandoned the base and left the animals. The animals were found almost starving to death in the small cages. The American forces called in special veterinarian to try to save them.

So, we had to pass those creepy empty cages to get some mango ice cream.

Before we went into the dining facility, we always had to clear our weapons. Without an ammo magazine in it, we placed the muzzle into a metal barrel, pulled back the charging handle one or two times, then turned the selector switch to semi, and squeezed the trigger.

Click.

This was to make sure we didn't have a round in the chamber and mistakenly fire a weapon in the dining facility. Did it happen that the round would go off into the barrel? Yes. And chances were that the soldier would get in big trouble for it.

Fabrizio was learning English so fast that he started making jokes, and he was so funny! I took a little pride in the fact I was teaching him to be very sarcastic and how to use all the good slang Army words in English. We were cracking jokes about an officer we knew, and one night at dinner he took it to the next level; I almost fell off my chair laughing. At this time a lot of Americans were getting kidnaped and the insurgents would post videos of them tied up as hostages. There was a catch phrase that this officer would say all the time, and Fabrizio acted out this officer being tied up on video and saying his phrase.

"We train to standard, not to time! Hooah Hooah! .. oh, and please help me," Fabrizio said while making a funny voice with his hands tied behind his back.

It was funny as hell, but also dark as hell!

Fabrizio, as your bad influence English teacher, I am impressed.

He would invite me to visit him in the International City to have some espresso in the Italian trailer. I thought it was pretty great that they had an espresso machine. But it was a little too soon after my dirty-Baghdad-water-brew incident, so I always said no. Also, if there was any small chance that he would have hit on me, it would have been in the Italian trailer with a tiny espresso cup pinched between his thumb and pointer finger. I just didn't want to put myself in that situation.

I don't know why I told him I couldn't ever go there. Lord knows I couldn't say both of those real reasons. I mean which is better? "Sorry, Fabrizio, I am still traumatized after my last coffee consumption." Or, "Sorry, you might put those Italian moves on me and then murder me when I say no."

There were not many female international soldiers there. However, there was one woman from Australia who really stood out. She was probably older than me, so maybe in her early thirties. She was really pretty with a slender build and curly blonde hair that she was allowed to wear down. That's actually probably why she stood out so much, too. The Australians also had form fitting uniforms that looked like pajamas to me. Fabrizio noticed her as well. He made some comment about her to me when we were together. I was like, "Dude, you are married." He told me that he could die here in combat at any time, so he should be able to live in the moment because it could be his last. A few days later he told me that he slept with her. I guess she said yes to have some espresso in his trailer.

LEAVING BAGHDAD AND THE LAST FEW NIGHTS

The countdown for me departing started. The new platoon leader arrived in Baghdad, and he was learning everything. He is going to be great, and I felt good about it. My platoon sergeant told me to come to the SSS tent at a certain time, and when I did the entire platoon was jammed in there for a surprise farewell for me! They got me a customized, embroidered guidon and a Baghdad plate that everyone had signed on the back. They got some extra 'near beer' and we told everyone that if they chugged a few near beers fast then they could get a buzz. Not sure if this was true, but it was fun trying.

That night meant so much to me. It is really hard to describe with words.

The Great Beer Run of 2004

Baghdad pretty much turned me into a shady character. Okay not really... but I did sharpen my skills in the art of bartering. I do not condone the actions I am about to relate. I honestly think it was a moment of insanity from being deployed, because I would normally never do any shady stuff like this in real life.

Since I was the platoon leader, I was the liaison between the technical contractors and my NCOs. My job was to interact with contractors and pass on the information to my commander and other leaders. In other words, I didn't fix anything or touch anything, I was just the messenger. But with this job, I made a lot of personal/working relationships. I made friends with three contractors, who apparently lived in a tower on the base. The tower looked really strange; it must have been an Iraqi look out or something. It was a thin tower along the palace lake, it had one small room, then a narrow staircase going up to the rest of the rooms. The guys said I always had an open invitation to come over for drink in their old Iraqi-tower-turned-house.

Alcohol was still nonexistent in Iraq because the country was dry. One of the only places you could get alcohol was from the international soldiers (the Brits) in Ballad, because I heard they had a bar set up. The other place was at the newly taken over Baghdad International Airport duty free shop. We weren't allowed to go there... but these contract guys were. Well, I had no desire to drink alcohol anyway, but the idea of getting it somehow was interesting to me because I learned fast how bartering was worked there.

I asked them if I could possibly have beer can. I was testing the water to see if they really had all this alcohol they were bragging about practically all the time. They came through and brought me a big can of beer. It was an English beer that came in a huge, long can. "Holy shit. I have beer in a combat zone. What dark path in life was I going down right now?" Anyway, I put this little golden egg in the back of my Gator, and I was on my way. I drove over to a huge tent where they were issued combat supplies to all of the soldiers. It was some actually very cool items, like new comfortable boots, black fleece, Under Armor shirts, sports bras, elbow pads, and other items. There was a guy there who was totally shady, and he talked about beer to me somehow in the past. So, I pulled up and we made shady eye contact and communicated without even speaking. This guy somehow telepathically read my mind and lifted the blanket in my Gator trailer and exposed the beer. How the hell did it know it was there!? So, the bartering began.

He wanted that beer. I said, "How about those Under Armor sports bras?" In all fairness they only issued us two, and the laundry situation was not the best. I always wore sports bras in the combat uniform. So more of them would have been helpful.

I was feeling gutsy. "For me and my other female soldier," I said.

Mr. Shady Dude didn't say anything and disappeared into the darkness. He came back with a box, placed it in the trailer, and took the beer can. There were about eight sports bras in there, and they sold for about at least $45 each at the time. I split them with the other female in my platoon.

So, the first drop went really well. I was getting confident in my shady ways. I was proud about our new collection of sports bras. The contractors did tell me to come to their tower and get more alcohol anytime I wanted. So, I started making my next plan.

It was the day before I was leaving Iraq for good, and it was time to put my plan in to play. I traveled to the tower on my Gator. (That kind-of sounds like a Disney fairytale, but I assure you it is far from it).

These guys were really technically smart...experts in their field sent here. Normally they were semi-talkative, but always nice and non-threatening. I knocked on the door, and they told me to come in. I walked in and immediately felt uncomfortable; I knew I shouldn't be there. They were all drunk. VERY drunk. Laying like a sloth-on-the-couch can't-move drunk. It was like an Iraqi tower frat party... minus the girls... minus the beer pong... minus the ... fun. There was a small fridge in the room, and one guy mumbled from the couch to take whatever I wanted. The fridge was filled to the brim with a lot of bottles of hard alcohol and a lot of cans of beer. Not a condiment in sight. I filled my cargo pockets with cans of beer, and I got the hell out of there. I left riding the Gator he gave me, and my pockets exploding with his beer cans. Yes, I was grateful, but I still wanted to get the hell out of there.

So, what was I going to do with all those beers? Well, it was my last night in Baghdad and as a thank you gift to my four squad leaders and my platoon sergeant, I wanted to give them a REAL beer. But little did I know I was about to have a heart attack on my Gator on the roads of Victory Base, Iraq.

I drove down the main road on my Gator in the pitch-dark night in the middle of nowhere Baghdad. Not many cars were on the small road normally, so especially not now. It was smooth sailing. As I was almost passing our brigade headquarters building, the only movement as far as the eye could see was the battalion commander (who liked me) and the battalion command sergeant major (who didn't like me) locking up the headquarters!

Oh, what the fuck!!

I didn't know what I was going to do. What were the chances that the moment I drove by was the very moment that they are coming out of the building? I had to get out of there. There was no way in hell that I could just zip by on my illegal Gator, all by myself, in the pitch black, with my cargo pockets exploding with items that clearly looked like fucking beer cans.

I reacted quickly and hastily took a right turn at the intersection.

Yes, the same intersection that the building was on. Where did that road lead to? Absolutely nowhere. It was a long dark road with just vast land on each side. Maybe at the end of the road was the base border? There were no buildings, and I cannot emphasis enough how pitch black it was. Also, my lights were off so I couldn't see shit. But still, there I was driving down this road because I knew that the other way was guaranteed not a good time, so I took my chances in the shadows of Baghdad. I was scared.

Also, I was thinking to myself... what the fuck am I doing!? And where the fuck am I?! How stupid am I right now, and God forbid something happens to me... everyone and their mother would say I deserved it for driving down this road by myself. I was scared of the enemy being there; but also, I was scared that they saw me make the turn and they were coming down the road, too. Or maybe they were standing at the intersection to wait for my dumb ass to come back.

I waited for a while. I didn't get killed by a terrorist, and I also didn't get caught by the command sergeant major (who didn't like me). I don't know which would have been worse honestly...

My heart slowed down from 300 beats per minute, so I cautiously drove back into "civilization." They were gone, thank God. I continued on my drive to the populated area of the base. My first stop was my platoon sergeant. I walked into the office trailer, looked him in his eyes, and told him to come outside. He stood up and followed as if he could literally read my mind that I was up to something, and it was going to be good. We went behind the trailer next to a big pile of trash. I reached in my pocket and said, "I got you some REAL beer! Thank you for being the best platoon sergeant ever." He looked at it and couldn't believe what I had and didn't even question where it came from; he was just glad to have it. I knew he liked beer, so I nodded like the Beer-Robin-Hood of Baghdad, and he knew I had more places to go. I let him enjoy his beer in silence while I walked back to my trusty steed.

My next stop was at the CHUs to the four other squad leaders. I wasn't as nervous anymore since leaving the danger zone of battalion leadership and company leadership. (What had I become?!) I knocked on the doors of my squad leaders; they were appreciative and played it cool to not draw attention. Hopefully they tactically destroyed the evidence (unlike the taco wrappers). It was late now, and I had to go back to my room to prepare to leave the next day.

R&R

Rest and relaxation! Doesn't that sound great? Deployed soldiers were all given two weeks for R&R, which meant we got to leave the war zone for two weeks. Soldiers were still charged for the leave, meaning that they used up the days they accrued all year; taking leave wasn't free. If someone didn't have enough days saved up, they went in the negative numbers, or as we called it, "going in the hole."

The leadership team made plans of who could take leave and when, so that not everyone was gone at one time in one section. Everything still must run smoothly while allowing soldiers to go on leave. Normally leaders would let all of their subordinate soldiers go on leave first. Since I was leaving Iraq to my new assignment in Kuwait, it made sense for me to take my R&R during the transition, rather than get to Kuwait then go on R&R in a few weeks.

My room was all packed up, and I only had a ruck sack and a duffel bag with me. I was able to mail a footlocker to my new mailing address in Kuwait. Those footlockers were the biggest boxes we had, and they were easy to get at the small PX. Everyone had a few of these lockers, and it was the main way we transported our personal items.

I had to turn in my M16 when I was leaving for the R&R shuttle. It felt really strange to give it away and have my arms and back empty. I felt lighter for sure, since I didn't have something heavy or pulling on my back and shoulder like I had for the past five months or so. A few days before we left, we had to meet with the group in charge of the shuttle and go over drills if we were attacked.

They transported us in a military troop carrier, which looked like an Army truck with the driver in the front cab, benches in the back, and big round roof and tarp for the cover. The higher ups were afraid that the enemies would catch on that this truck always had a back filled with unarmed soldiers. So, they would change up the times of departure and take different trucks with different bumper numbers. There was an armed soldier in the passenger seat, and one in the back with us sitting ducks. At the rehearsals we were told, "If we get attacked, drop down on floor of the truck and cover your heads." Simple enough... I can do that. Stop, drop, don't roll. Got it.

So, I was leaving Victory Base. Leaving for the first time (officially... shhhh!) and hopefully never ever returning again. I didn't feel ecstatic. I felt strange. I had guilt for leaving early before my platoon was able to. I felt bad that they had to stay in this danger zone, and I was going to Kuwait where there was little to no threat. I felt sad for the soldiers I knew who died on or around the base while I was there, and how they didn't get to leave like I was. I surely didn't want to stay, but I felt like there was unfinished business still and I was leaving my team, even though I knew they were in excellent hands with the new platoon leader.

The tarp flap closed, and there we all sat in the darkness just looking at each other. We didn't want to look at each other, but we were literally face-to-face, so we had no choice! We were just sitting there on high alert, listening to the sounds of the road, and feeling every turn; it was kind of like being on an airplane and listening to the landing gear unfold to make sure it sounded normal... even with zero flight knowledge whatsoever.

I remember thinking to myself, "Please God get me to BIAP. I didn't survive these months in Baghdad to die sitting defenseless in this cattle car." The only way I could fight back the enemy would have been with verbal insults, and I don't think that would have gotten me very far. Anyway, what was I supposed to do again? Drop, stop, jump, run? I forgot.

We arrived at BIAP safe and sound. I started what was going to be a very long "day," although I didn't know it yet. After a long time of sitting and waiting, which I most likely blocked out of my memory, they brought us out to the runway to board a military aircraft, a Chinook helicopter. I remember very clearly the moment my boot lifted off the cement runway. I am *really* leaving Iraq. How dramatic am I? But I was really looking forward to not having my boots on this ground, and never coming back ever again.

Lift off.

There was a bad ass soldier standing in the open doorway during take-off with his weapon pointing out the window. He was hanging almost out of the freaking door, looking left and right. Then when we got high enough and out of danger, he closed the door. Thanks bro. What a job! Coming from the signal communications world for so long, I stared in awe at this badassery that stood in front of me.

We landed at the airport in Kuwait, but in a portion that was for the US Army. It was hot and there was sand everywhere. It was a tent city again.. Charming. They assigned us to a hot transition tent filled with cots. This is where we could stay until it was time for our flight. We were able to get dressed in civilian clothes. This is the first time I wore civilian clothes for about five months. Our packing list had us directed to include two civilian outfits. This is one of the instances that we had to wear civilian clothes.

What kind of outfit should I pack for combat?

Yeah, that was something I never thought I would have to consider... but it was on the packing list! I packed two capri pants from Aéropostale - one blue and one kaki. Then two plain scoop neck tops. Totally appropriate. I declared a new dress code category: Combat-casual.

I left my bag on the cot and hoped for the best that someone didn't steal all my stuff and then fly off to only God knows where with it. I went to a tent where a woman was sitting at a desk with a computer and printer. Two men were in there already, so I sat in the waiting room part of the tent... aka the metal chair next to them.

They were two contractors who were also coming from Baghdad

and were also going on R&R. Their conversation went something like this:

"Where do you want to go? Somewhere warm right?"

"We could go scuba diving."

"What about Australia?"

"Okay."

They both simultaneously turned to the less than enthusiastic travel agent, and said, "Sidney."

How many times in your life do you get the chance to pick a flight to ANYWHERE you want in the world for free? Yes, for free. R&R flights were free to anywhere you wanted to go. I have thought so many times about this moment, whether or not I made the right choice. The sky was the limit... literally! I have never been to the west coast before - California would be awesome. I could steal these guys' idea, go to Australia, and see some kangaroos. Hawaii? Because why not? It looks like a great time, and I could dance with a grass skirt and coconut bra.

However, while it was not very exciting, I went to my home in Germany. I was tired. I was also alone, so would I really want to go to these amazing places by myself? I wanted my bed and my shower. In hindsight, I should have gone home to New Jersey to see my family, but at that moment I just really wanted to sleep and relax.

My flight was later that evening, and we flew on a contracted plane to Amsterdam. There were no other people but military on the plane. I was in the middle seat and next to me was a really large and sweaty male contractor. I was so tired at this point that I didn't even realize that I fell asleep right away when the wheels went up. I woke up when we were almost there, and my head was resting on his shoulder. Oh, it looks like I missed dinner, too. This is awkward.

I arrived in Amsterdam and found out I had an eight-hour layover in the airport. I could drive home in four hours. I was so frustrated and tired, I cried when the KLM lady printed out my ticket. I said, "I just want to go home."

I had been up since the morning that I left Baghdad; I was tired.

My only power nap was on my seat neighbor's big sweaty arm. Now I had to stay occupied for eight hours. I could have gone into the city, but I didn't want to risk not getting my flight again, and I was tired as anything and didn't have a desire to look at Dutch master's paintings and graffiti of blunts with faces at all the smokes shops. I really just wanted to go home.

I kept myself occupied in the airport by looking at shops and buying make-up. I hadn't worn make-up in months and that was for sure not on my packing list. It felt like a weird dream. Did I really wake up in Baghdad this morning in a combat zone, and now I am comparing eyeshadow pallets in Amsterdam?!

My hour-flight to Frankfurt went by fast. I was most likely contorting my eyes and face the whole time anyway with my make-up haul. I took a taxi to my house, and I didn't give a crap about how expensive it was.

I walked into my apartment, dropped onto my bed and fell into a deep sleep, where for the first time in a long time.. I felt safe.

Since I didn't prepare so well before I left for the deployment, I had the feeling that I wanted to go to my apartment. I should have gone home to New Jersey and visited my grandparents while they were still alive. I should have gone home and spent time with my Mom, who I now miss every moment of my current life. However, I went back to Germany and went on some trips to include a bus trip to Paris. R&R is two weeks, and my Dad came for one week, which ended up being nice.

My two weeks went by fast. I did some tasks on the base that I had to get done. I remember having to wait in long lines doing some things, and I felt like I should have been bumped to the front because my time was ticking to go back downrange.

It felt SO good to be home safe and in my own bed, bathroom, and space. It was a feeling that I never felt before and can't describe how good it felt. It was a real appreciation for the "normal" things in life. Like P. McCree Thronton said, "To those who have fought for it, freedom has a flavor that the protected will never know."

Then time was up, I packed up some more civilian workout clothes since I didn't have any, and I took a commercial flight to Kuwait City. The airport looked like a fancy shopping mall made in marble. All of the advertisements looked the same, except the women were all covered in burkas. On my flight there was a white girl around my age, with long, pretty platinum blonde, glamourous hair. While we were collecting our bags, she was wrangling up her massive stack of Louis Vuitton luggage bags. I have never seen anything like it. It sure blew my knock-off Louis Wuitton from Canal Street in NYC out of the water. Who was this girl and why the hell does she live here? With some rich Kuwaiti man? In some penthouse apartment in Kuwait City with a big Louis Vuitton couch? In my version of her life that I just made in my head as I stared at her at the luggage belt...she does.

I got my non-Louis Vuitton green duffle bags and headed to the area where unit representatives picked up service members to take them to the middle of the desert for the next few months of their lives. I hadn't been to Camp Udari since the beginning of the year. The camp was a staging post for troops heading into Iraq. The area surrounding it was known as the Udari Range Complex. No one lived there, but later I learned that was uninhabited except for a few nomadic Bedouin tribes raising camels, goats, and sheep. Sounds wonderful doesn't it?

By the time I came back they had changed the name to Camp Buehring in honor of Lieutenant Colonel Charles H. Buehring who, in 2003, was one of the highest-ranking casualties of the Iraq war. He died during a rocket-propelled grenade attack on the Rasheed Hotel in Baghdad.

A senior NCO picked me up from the airport to drive me to the camp, and he filled me in with all the unit changes since I left. And more importantly... all the gossip.

Kuwait, Take 2

Permanent Party Peeps

When I arrived in Kuwait, I was now the executive officer (XO) and also the acting commander to about 75% of our company who were still in Kuwait and not in Iraq. I had a big learning curve, but I was ready for the challenge. My mission and daily tasks would be very different than in Iraq, but the overall mission of taking care of my soldiers stayed the same. Since I left Kuwait, a lot of aspects of the living situation changed.

While I was in Iraq, our unit in Kuwait moved out of the group tents and upgraded to big trailers. I guess you could say we were in a trailer park. Each trailer was very large and long with a hallway down the middle and two big bedrooms on one side, and two big bedrooms on the other. Of course, all females, males, officers, and enlisted were separated at this point. In each room were six bunk beds, so there were twelve people in each room.

There was a MASSIVE air conditioning system attached to each trailer. Each room had a big hole in the wall and a tube would blow freezing cold air in the room. I mean freezing cold. We would have to block it up with a pillow, to make it less cold, and the people on the bunk right next to the hole would freeze at night.

The bunk beds were made of round metal poles, to include the poles on the stepladder. Walking up the ladder was extremely painful on the feet. I had to make up different ways to climb up the ladder because going up normally was like some kind of Chinese torture. I was the newbie coming in from Iraq, and the gals there had been settled for about six months; so, when I got there I had to be on the top bunk. Also, the top bunk was for the lowest ranking person. I am guessing this is also how it works in prison. (Actually, I am not guessing, I watch a lot of prison documentaries on Netflix.) So that's fine, up the death ladder I went.

After a while, a captain moved on to take another job somewhere else because her command time was over. So, that meant the first night home-girl was gone, the newbie moved on over to her spot. And when the next newbie came, they had to go on the top bunk. (Again, pretty sure same deal in prison.) Then I was living large. A mattress on my spring bed, no one above me, a whole metal bunk to myself. I could even hang a sheet down from the top to make a little bunk bedroom. The lap of luxury for downrange standards!

There were no bathrooms inside, so we had to use the port-a-john outside the trailer. At night the scorpions would find refuge and warmth inside them, and many people got stung in the middle of the night. Needless to say, I trained my bladder to be like iron, and I did not let myself go to the bathroom in the middle of the night. Yes, this was uncomfortable but necessary. It's another reason why you should never take your private bathroom in your house for granted.

Taking showers in Kuwait was a big effort, but also a good time killer, and we had nothing but time to kill. The showers trailers were on the backside of the trailer city that we lived in. Everyone had a little basket filled with their shower items. We had to get dressed in our Army-issued gym clothes and shower shoes, then walk to the showers across the sand. Sometimes there would be sandstorms and right after we got out of the showers, we would walk right into a wall of sand in the air. Then it would get all stuck to any wet skin and hair again. Just wonderful...

Also, all showers had to be "combat showers" which meant that we would have to lather our bodies and hair with the water off. Then we could only put the water on to rinse, and it couldn't be for a long time at all. If someone heard a shower running for a long time, others would yell, "Combat shower!!" This was to save hot (or warm) water for everyone. The water came from big tanks right outside of the shower trailers, and when the warm water was gone... it was gone.

Looking back, it seems kind of strange and unnecessary, but there was a young girl who worked in the shower trailer. Her job was to keep it clean, I guess. So, we would carry our shower baskets into the trailer, and she would be sitting in there. She was really nice and friendly; however, she didn't speak any English. Maybe she learned a few words being in there like hello and goodbye. Even though she didn't speak English, we all talked to her and tried to communicate.

She was from Bangladesh. I didn't know anything about Bangladesh except that a lot of H&M clothes were made there. She was really young, seemed to be from a poor family, and didn't know much about the world. She would get super excited about everything we showed her. Everyone would bring her some gifts and let her have the leftovers of our body wash. She would always smell the different scents of all the new kinds of body wash and really like them.

I talked to her every time I was in there. It seemed like she understood everything we were saying, but she just couldn't speak much English. One day were communicating with drawings and hand gestures, and she posed a question to me. She pointed to a drawing of America and then to me to ask where I was from in America, and I told her I was from New Jersey, right near New York City. She didn't understand. I told her where the World Trade Center was located before 9-11. She had no idea what I was talking about. I tried to tell her how it was a part of the whole reason we were deployed here. Still nothing. So, there I was in a desert in the middle of nowhere Kuwait, pretending that my shampoo and conditioner bottles were the World Trade Center towers and that my body wash bottle was the hijacked aircraft flying into my coconut Suave twin towers. Still... no clue.

I mean, how would she know? I had seen people living in clay huts in the middle of the desert in Iraq. Maybe she came from a really poor area, and they don't have TVs or newspapers. What did she think we were all there for anyway? Also, how many other people there didn't understand what we were doing in their country? These were deep questions for me and too much to handle at the moment. I had to focus on my quick combat shower.

Life at Camp Buehring was like the movie Groundhog Day: the same routine every single day, and we had to keep going through the motions. In Iraq there was a lot more people, buildings, money, and places to get into some mischief if you were bored, but in Kuwait... not so much. It was like Mad Max living - people making stuff out of junk they found and pimping out their sleeping areas to make little rooms with curtains.

I would pass the days with the same routines. My supply sergeant and I would go to aerobics that was taught in a tent every day. It was a lot of fun, and we looked forward to it. It was taught by another soldier who was stationed there as well. She had the same CD playing each time, and somehow it was great. We didn't have music! This was before any streaming of anything or buying online music really. So, her one CD was fire. One day a track got scratched and she had to skip it. We were devastated. (Damn that was my jam...)

They set up a gym in another tent toward the end of our deployment, but it was extremely crowded and kind of shitty, so we never went. We did go in there to get some water at times. They would drop huge pallets of water in front of most big tents, and it would sit out in the sun for days. The water would be extremely warm and almost hot when we drank it. Put some tea powder in there and you have a big plastic bottle of warm tea. And we drank it.

WE HAD no weapons on us! Deployed, in a combat zone, with no weapons! They were all locked up in the arms room. This was very strange for me, coming from Iraq and having my weapon on me 24-7. I literally slept with my M16 in my bed next to me for months. If shit went down while you were 'weaponless-in-Kuwait,' the armor NCO was supposed to run to the arms room, and I guess us, too, and we had to get our M16s. Not such a promising scenario, but that was the plan. There was always a quick reaction force (QRF) on duty on the camp somewhere. The QRF consisted of just regular permanent party soldiers, and no one was actually trained to be a force protection security team.

We were the permanent party there now, and everyone else was using it as a staging area before going up north to Iraq. When soldiers found out that we were there permanently, they felt bad for us. They were heading into harm's way to fight terrorists, yet they pitied us. A lot of infantry troops passed through, and they were a different type of breed than I was used to being around. For example, overhearing them talk about using camels as target practice, serious cursing and dirty talking, more than the normal Army cursing. And of course, the disgusting artwork that would decorate the inside of the port-a-johns when they left. I think if a physiatrist did an analysis of all that graffiti, they would be mortified.

I was twenty-four years old (turning twenty-five toward the end of the deployment) and I was in charge of soldiers who were about four years younger than me; many joined right after high school, but I went to college first. Also, I was in charge of grown men and women who had already been serving for almost twenty years. For the most part this all worked out fine. They respected me because of my position and rank, and I respected them for their rank and experience. They knew that I understood that they knew everything, but they still respected me as the commander.

While this worked fine when it came to Army business, when it came to getting in trouble and acting out, I found myself in situations thinking, "How are you a grown man and I am here, twenty-four, telling you how to act right?" One of our older mechanics was told that his R&R leave to Germany was postponed by a week or so, and he flipped out. He got so pissed that he ran out of the OPS tent and kicked the door open to get out. He kicked it open like in the movies when a SWAT team is about to clear a room. I couldn't believe what I was looking at, and in my head I thought, "Dude, we MADE that door! Take it easy!"

I followed him to tell him to calm down. I believe we needed him to stay to ensure some vehicles were serviced. It wasn't my call to keep him there, and I most likely wouldn't have made him stay. But that was the way it was, and we could not change it. I couldn't believe that it was my job to talk to a grown man like this.

He most likely had some German girlfriend waiting for him to come home, so that is why he was so mad at the change. The only reason why I feel confident to jump to this radical conclusion is because he used my computer at nighttime apparently. I was on my laptop trying to send a PowerPoint slide to the battalion, and when I clicked search to find the file, I saw photos of a woman.

"What the hell?! Who in the hell?!"

I looked in my documents and found a whole bunch of photos of an older German lady with bright red hair who took photos of herself standing in her bathtub with lingerie on. This was before smart phones and selfies so she must have propped up her camera on the sink and hit that ten-second timer, so she could run in the bathtub and strike a pose. There was no nudity, but it was still ridiculous. I guess when he clicked download from his AOL or Yahoo email, he didn't know it was going to save to the main hard drive of the laptop. Computer security was not what it is today obviously. I just thank the good Lord that I didn't see any of the photos that he was sending her.

The staff sergeant who worked under him in the motor pool also used someone's laptop at one point and made a computer oopsie with me, too. My commander wrote to him asking about some inspection of the vehicles, and since I was the executive officer, I was in charge of the motor pool, too. The commander in Iraq wrote asking about it and copied a bunch of people in the email. The staff sergeant replied to all, but he didn't realize that he did. His reply to the commander's question included, "Well if LT gets off her ass, and comes down to the motor pool she can come see for herself." I laughed and thought it was hilarious. I still do. I took the hint, and I got off my ass and went down there.

It's Lonely at the Top

As a people pleaser and a social butterfly, it was a hard lesson for me to learn that "it's lonely at the top." One of the first hard lesson I learned in the Army was that not everyone is going to like your decisions or even like you in general. This may sound like it's not a big deal, but at the time I was coming from being a platoon leader with platoon sergeants who shielded me from any type of situations like that. If it was a good cop / bad cop type situation, they for sure took the bad cop spot for me.

Also, there is a fine line between wanting to be liked and cool with the soldiers, and then allowing things you know are not acceptable because you fear to lose the "cool officer" status.

We only had one phone line in the company tent, which sat either on my desk or the first sergeant's desk. Our desks were right next to each other, so this was not hard. One day I was on my laptop most likely doing PowerPoint sides, and a staff sergeant squad leader, who was a very nice, cool guy, sat next to me to call one of our sergeants first class who was on leave in Germany at the time. He asked the sergeant first class to bring some "HenHen" back with him. I'm not sure if he was afraid that CID, the FBI, or ISIS was listening to our phonelines, but even a young lieutenant, literally so close our knees were touching, could understand that he meant Hennesey.

So, what am I to do? He just blatantly said right in front of the commander (me) who was in charge of the safety the whole company, that he was going to bring in an illegal item into the camp. And just hypothetically speaking, if I was to stay as the "cool officer" and not say anything, and he did get caught by someone else, like the command sergeant major, and said, "Well LT knew we had it and she didn't say anything about it, so I thought it was okay," what kind of pickle would I be in?

So, there I am, stuck on this fence-like situation, wondering if I just turn a blind eye or do something about it. Looking back, I wonder, and maybe you wonder, why didn't I just say something in the moment? Why didn't I grab the phone and in a joking way say, "Oh the hell you aren't bringing that here!" It was because I was in shock. I was also too young to react so quick to something like that, like I would react now. Also like I said, at this point I never had to be the "bad guy" before.

So, I consulted First Sergeant Minty and told him what happened; he was surprised. He told me he would take care of it. So, I left it alone.

A few days after that I heard we had a company health and welfare inspection at around 0300. They didn't wake me up, and I respect that they wanted to fix any issues within the NCOs and not get the officers involved. What this inspection entailed was everyone waking up and going outside, while the company leadership went through everyone's stuff, looking for items at were not allowed downrange and against General Order Number 1, which prohibited certain activities for soldiers, and promoted good order and discipline, including no alcohol; everyone was treated the same in the check and no one was targeted. Also, they weren't allowed to rummage around, they were only allowed to see what was in open view. Before they checked, individuals were all also asked to bring any contraband items forward ahead of time, so they wouldn't get in trouble. It was called an amnesty box, and no questions were asked.

Did this happen that morning? I don't know. Did they find more alcohol bottles to include the requested alcohol? I don't know. What I do know is that the first sergeant did take care of it before the battalion did a check and weren't so nice about it.

The staff sergeant knew I told on him, and chances were that the rest of his alcohol was taken, if he had it. He hated me afterward and basically ignored me for the rest of my time there. He probably still hates me to this day. That stung a little for me because he was a really cool and funny NCO, but like I said, I did learn what it means when people say, "It's lonely at the top."

TRIPS TO CAMP DOHA

We would go to Camp Doha almost once a week for either a morale trip or some fake business we had to conduct. The feeling of freedom was incredible because we were actually leaving the base and could see something new. Also, we didn't feel so suffocated as we did on Udari, which was literally like a sad camp in the middle of the nowhere desert. When we went to Doha we had to have our weapons with us, and we dressed in full battle rattle. It was such a strange comparison to the Kuwaitis driving around.

Once we were close to the entrance of Doha, I was in the M998 Humvee in the passenger seat with my driver, and this fancy sports car with a bunch of young guys pulled up next to us. Since I was just back from Iraq and clearly still deployed, my fight or flight awareness / sensors went sky high. They were checking out the young blonde lieutenant, and I was just hoping they didn't shoot up our Humvee. We were passing a McDonalds, and to me, coming from Baghdad and now living way out in the desert, that was also a culture shock to see that there were "normal" people living "normal" lives taking their kids for Happy Meals.

Right when we drove into Doha, there was a small conex building – made from large metal shipping containers - where some "DVD homies" ran the shadiest DVD shop around. When you walked in they would talk real cool and everyone would laugh. They would say things like, "What's up my homie?" and "Fo sure my dawg." Everything was bootleg, burned and reburned from original movies, or they were bootlegs from filming in the theaters. It was a gamble if they would even work, but we were desperate, and the price couldn't be beat. You could bring them back next time and swap them if they didn't work, but it was still annoying. How the military base allowed these guys to have a shop was seriously beyond me.

Doha had a legit food court with so many options, inside an actual big building. You could even get Baskin Robbins there! How unfair was that for us out in the middle of nowhere? They had a small but real PX and a lot of different vendors with stalls. I guess they were taking advantage of everyone receiving combat pay and killing boredom with shopping.

There was a morale building that they called the Crystal Palace. People who were stationed in Doha would actually go there for a relaxing day... as if their regular living situation wasn't relaxing enough. People would sunbath in the grass and swim in the nice pool they had there. Female soldiers wore bikinis but only for a short time because word (or photos) got back to the spouses at home, who complained to leadership, then it was not allowed anymore. The Palace also had a big meeting room and also a music room with all types of instruments. I thought the music room was pretty awesome and such a good idea for downrange.

DON'T PISS OFF YOUR DENTIST

I have to set the stage a bit as to where I was in my life at this point when it comes to my attitude. I was the acting commander of my company, around Army men all day, just returned from the warzone in Baghdad and – remember - still a Jersey girl. So, wrap all of that up together: I didn't take shit from anyone and was not afraid to let them know it.

While I was deployed, my wisdom teeth were starting to hurt. They would throb one day and then be fine the next. I didn't know if I wanted to get them out in Kuwait or not, but I felt like I had to at least have them looked at. I hadn't been to a doctor or dentist since I deployed. The medical personnel in Kuwait were all Reservists. Therefore, they were not used to being on active duty all the time (in the real world they had civilian jobs), and they were normally much older in their ranks. Also, the dental clinic was in some trailers lined up in a square. I made an appointment and was ready to go to it.

In Iraq we were only allowed to show up to the medical facility in the physical training (PT) uniform. So, now in Kuwait, I got changed in my PTs, and off I went. The PT uniform consisted of black windbreaker shorts and a grey t-shirt. We could add layers for warmth, but I didn't need that in the desert.

I walked into the trailer and the dentist, or the dental tech, was a really old Reserve staff sergeant. He was tall and his hair was fluffy and out of regs. The minute I walked in he was already rude to me, and I immediately noticed. I didn't realize yet he was being rude to me because he thought I was a private or specialist. I looked really young, and I didn't have any rank on my uniform. He gets on to me about me wearing my PT uniform to the appointment and how unhygienic it is to be wearing it in there and about all the dirt and sweat that I was dragging in. (My uniform was obviously clean, and I didn't work out in it). I didn't know why he couldn't just tell me in a normal or professional way. I explained to him calmly, "Sorry in Baghdad, we had to wear PTs to our medical appointments."

Then the conversation was about the potential wisdom teeth pulling. He told me that the recovery would be hard, and it wasn't the best in this environment here with the sand. When he said here, I thought it meant on our base because it is so sandy. So, my next question was, "Should I go to Doha (the bigger base in Kuwait) instead for the wisdom teeth?"

Then he says to me, "Why? Because you think they will let you wear your PTs there?"

That's it dude, I am done being nice.

I said to him in a very excited high tone of voice, "Ohhhhhh!! Woowwww... you are soooo funny! Are you a stand-up comedian dentist?" (Come on, that's funny...)

His eyes got huge! "What did you say? I am going call your first sergeant!"

(Now it was confirmed with me that he did think I was a young soldier.)

I said, "I don't care... please do!!!"

Then he said, "I am going to call your commander!!"

I said, "...... I AM THE COMMANDER!!!"

Mic drop. And I walked out.

Needless to say, I just delt with my teeth pain for the remainder of the deployment and didn't let that ass hat put his fingers in my mouth.

LIFE IN THE MIDDLE OF THE DESERT

Since we were tax free and getting combat pay, we were all making the most money we ever had made. With no online shopping really back then, the only places to shop were on base. On Udari we had some trailers with vendors and sometimes there would be a bazaar that would visit for a week or so.

The shops in the trailers had very strange random things to buy. It was mostly middle eastern decorations and trinkets. There were a lot of rugs, things made of marble, and Arabic painted wooden boxes. I didn't buy many things from the shops. I wasn't really interested in that type of décor and also there was always the problem of getting it home.

No doubt we still went into the shops a lot just out of boredom. The workers there were a mix of Kuwaiti and Indian. There was one man I would talk to and chat with... just being friendly. I told him I was vegetarian and I think that made him have a love connection with me. Once he showed us the most messed up desk toy ever possible. It was a stapler, and when you push down on it, the thing would electro zap the palm of your hand. It hurt!! I mean it hurt so bad it made me feel nauseous. Why anyone would buy that is beyond me. A totally messed up gag on a co-worker and, considering the number of volts in the zap, I am pretty sure it was illegal in America. He also had click pens that did the same. I did not try them...

So we went in another day, and I started talking to another Indian man working in the shop. I didn't think twice about it obviously. The original guy I talked to before got REALLY mad at me. He was shouting things like, "Oh you talk to him now! Why are you talking to him?" He got crazy jealous. In his mind I belonged to him, or we were in some type of relationship. I was shocked about it all. I pretended I couldn't hear him from across the store. I took my last look at the random items for sale and got the hell out of there, and never went back. I remember being scared he would find out where we slept on the base and come find me. (Damn maybe I did need that pen zapper after all...)

So those shops were off limits for me now due to my break-up with my Indian boyfriend I didn't know I had. eBay was a thing then, so I would bid on the most ridiculous stuff out of boredom and the need for some excitement in a few weeks when I received the package.

I bid on a Von Dutch hat on eBay that was way over-priced and most likely the guy was bidding against me from another account. I ended up paying about $70 for it. When it came in the mail, it was pretty obvious that it was a knock off. I felt so stupid, and I was pissed that someone could rip off a person who was downrange in the military. Yeah... but I still wore the hat.

To our surprise, fast food trailers started popping up and they formed a square near the shopping area we already had (right next to my Indian fiancé's shop). There was now a Subway sandwich shop, a Green Beans coffee shop, a mini-Burger King on the back of a truck, and a spot for new vendors to sell items.

My favorite vendor was the guy with the big fuzzy animal print blankets that were sold in plastic suitcases with handles. I bought and mailed some for friends back home. I wish I had gotten more for myself and sent them home. I wish I had a different one each day to wrap myself up in like a fuzzy burrito.

Subway was like a godsend for me because the Army dining facility food was, first of all, terrible, and second, had hardly any vegetarian options. I would normally only go to the DFAC for lunch because the long walk with sandstorms was just too much effort to do it twice in a day. I would skip dinner and have some crackers in the tent or something. However, when Subway came everyone, especially me, would go there for one meal a day for sure. I would always get the same sandwich: the veggie sub. The Kuwaiti workers at Subway liked me a lot. Maybe because I am friendly and nice or maybe because I was one of the few females! Regardless, I got special treatment, and for all the other times in the Army when I get the short end of the stick for being a female, I took any Subway perks. They knew exactly what sandwhich I wanted each time, so I didn't have to order it. Also, they had a promotion where you got a stamp each time you ordered a sandwich, and when you filled your card you got a free six-inch sub. Well, each time I went in and bought a sandwich they would make it rain stamps my way. Rip! Ripped that stamp reel each time and gave me about ten stamps. So, for every sandwich I paid for, I had about one or two free ones coming my way.

After months of eating Subway every day and the same sandwich, when I came back to home to Germany I ... could... not... eat... it!!! No matter how hard I tried. It made me think of Kuwait, I was right back there. The taste made me sick. It took me about three years before I was able to get a Subway sandwich again.

Not to toot my own fast-food horn, but I was also popular in Green Beans. For some reason I didn't drink coffee at this point in life. I didn't like the taste, and I didn't want to have withdrawal if I didn't drink any. I was too young to appreciate the awesomeness of coffee. Now I would have it injected intravenously all day if I could. However, in Kuwait, I would go to Green Beans Coffee for the excitement of going somewhere and feeling kind of normal again at a café. I would order a chai latte every single time. Again, I didn't need to order anything verbally. I could flick my pointer finger up for a split second and my coffee dude would be already frothing that milk. They called me "Chai Latte." I would go get coffee often with the same person from my unit, and when he went in there without me, they would ask him, "Where's Chai Latte?" and he would put his hand out flat by the middle of his chest, meaning my height. Damn, I didn't know I appeared like an oompa loompa to most people.

. . .

THE LAUNDRY SERVICE on Udari had a lot to be desired. When I had dirty clothes, I had to put it in a laundry bag and bring it to the laundry tent. It was like Iraq but not as nice. When I got my laundry back it all smelled like it was soaked in skunky armpit water. So after a couple of times dropping it off, I quickly got creative on how to wash my clothes myself. Some people got buckets and washed their own clothes by the sleeping trailers. It was not uncommon to walk outside at night and see people hanging their clothes on chairs or spinning their pants around their heads like a crazy raver at a concert to dry them.

We had a small PX toward the end of the deployment. I went there to buy a lot of post cards to send home to people. It wasn't really a PX, it was more of a snack shop, with so many chips, sodas, nuts, and magazines. I wasn't into eating snacks, so I hardly went there. The first time I went, I had civilian workout clothes on for some reason. One the NCOs in my battalion did a double take and then came over to me. He said, "Oh my God LT, I saw you and I thought you were someone's teen child! I was wondering how in the world this was an assignment where you can bring family members?" Back then I was embarrassed, but now obviously it would be a huge compliment!

MWR MONEY

I have always had a talent of acquiring things or figuring out how to get stuff for my soldiers. You can acquire items by a "Night OPS" situation, where you steal it... but in an Army way: strategic transfer of equipment to alternate location (STEAL). Or by finding out ways of financial support and programs that are available So, when I heard that there were MWR funds available somewhere somehow, I was all over it like white on rice.

Through the grapevine, I was told what office to go to in Doha to fill out the required paperwork. They did not make it an easy process to say the least. Apparently, commanders were allowed a certain amount of morale welfare and recreation money per soldier per month. It was like a strange soldier child support, and I was ready for it.

I drove the hour to Doha and found the MWR office in the middle of a parking lot in the back of the base. The office was freezing cold from an air conditioner, and it reeked like that stinky Kuwait AC armpit smell. I filled out the paperwork, and they cut me a check for the soldiers in my company. I got a few hundred dollars. I was allowed to spend the money on food, but it had to be tied to training as well. You had me at food.

I used my one laptop in the company which had internet running to it, and I found a restaurant in Kuwait City that would make us food and deliver it. They were able to make different theme buffets for us, for example, Tex Mex, Indian, and BBQ. This was a big deal and since it was shared money, I put it to a vote. I listed the menus of each buffet and put it on the wall of the OPS tent, and everyone voted for the top themes.

We reserved a tent near our area and had our feast ready to go. One of my soldiers took a Gator three-wheeler, went out to the front gate, and collected the food from the restaurant guy. This was the strangest Uber-eats type transaction I ever orchestrated, but it worked. Staff Sergeant B came down the long dusty road from the front gate with our Kuwaiti Tex Mex BBQ in tow. We all ate like we were kings in the paintings where everyone is jolly, pouring wine everywhere, and eating grapes from the bunches (minus the wine). After nine months of having bland terrible food, this was awesome. We ate every last grain of rice. After we ate, we had some annual mandatory training, because I had to stay in line with the funding requirements. A jealous company commander stuck her head in to see what we were doing and if we were really conducting the training. Lame. We had a dinner like that once a month for the last three months of the deployment. I used every penny we were granted.

Birthdays Downrange

As the company commander, and also as a person, I made a big deal about people's birthdays downrange. We were away from our loved ones, and it felt really terrible having another Groundhog Day on your birthday.

I made friends with the Indian workers at the dining facility, and they would give me whole cakes. I would pick up the very nice, boxed cakes at the back of the DFAC and then gift them to my soldiers on their birthdays.

A highlight on the birthday calendar was First Sergeant M, who was was an awesome guy and everyone loved him. He was a character - a motorcycle enthusiast from Boston who loved the Red Sox. As a Yankee fan, I talked a lot of trash during the 2004 World Series, so it was pretty lively when I walked into the OPS tent the morning after the Red Sox won after so many years.

When his birthday came around, I knew I had to make it very special. I planned the day like a fine theatrical masterpiece. To start, I was careful not say anything to First Sergeant M all day long. ALL DAY. I told everyone to NOT say happy birthday to him. He always packed up his bag at the end of each day and went to him room to watch a movie. That day, I said goodbye and good night like any other day, while it was killing me inside, knowing that it was his birthday. You see, if I let him know that I knew it was his birthday that day, the prank would never work, because he wouldn't believe anything out of the ordinary. But now it was showtime.

Everyone, take your places.

The first step was for all of the soldiers to get in the OPS tent - all ninety of them. They were like sardines in a can. Then I sent one of our gentle-giant, calm soldiers on our Gator four-wheeler to First Sergeant M's tent to tell him that there was a problem at the OPS tent, that the MPs were there, and they wanted to talk to him. He told First Sergeant M that I and the other first sergeant who, came back from Iraq, had and altercation. He knew that I and the other First Sergeant didn't get along, so this was semi-believable. He was shocked! He said later that the first thing he thought of was, "Shit, I missed it!"

The soldier drove First Sergeant M to the OPS tent, but before he could go in, I had the battalion command sergeant major outside the tent a few feet away. He stopped him and said, "First Sergeant M, how could you have let this happen?! The MPs are in the tent, and they want to talk to you." First Sergeant M was shocked, and his face turned white.

Then he got to the OPS tent door, and I had the highest-ranking officer, our battalion commander, Lieutenant Colonel T there. She stopped him and said, "First Sergeant M, I am so disappointed in you, I thought you would have warned me something like this would happen." Now he turned white as a ghost.

He walked in with the command sergeant major and the battalion commander to a jam-packed OPS tent, and we all yelled, "You got Punk'd!!!" (The prank show Punk'd by Ashton Kutcher was really popular at the time.)

When he realized that he wasn't there for some military police interrogation, his face had a huge wave of relief and said something along the lines of, "Holy shit!"

Then we all sang happy birthday to him and had cake, like I wanted to do since the morning! First Sergeant M was an awesome NCO and overall person, so he deserved a good downrange birthday... and prank!

STARSTRUCK IN KUWAIT

Besides pranking people, there was other entertainment possible in Kuwait. Many famous people came on tours to downrange locations, which were hosted by the USO. When you think of these tours people automatically think about Dick Clark and Marylyn Monroe on stage surrounded by adoring troops. The actual visits were meet and greets where soldiers stood in line and, one at a time, the celebrities' hands and say hello. Sometimes the soldiers could even take photos with them. Most of the time, especially back in those days, the celebrity didn't get any photo ops or fame from coming to visit... it was completely out of the news.

The first visit was from Henry Rollins. How cool! There was no way we were missing that. He came all the way to our base way out in the desert of Kuwait. We got to stand in line and meet him.

I was standing in line with a few of my soldiers from my company and a few of them asked Henry Rollins, "So if you are in a bar fight, what is your finishing move?" Valid question.

After my guys had taken their photo, they asked him, "Can you give our LT a kiss?" While I was grateful for my guys looking out for me, I was kind of embarrassed and laughed it off. I said to Henry Rollins, "Thank you for coming, we really appreciate your support!"

Then we were given photos with his autograph on it. I still have it hanging in my office today. Fast-forward fifteen years later, and he performed at the Wacken metal festival – the largest in the world with 80,000 people. I didn't find him, but I really wanted to and tell him that I met him in a tent in a small sandy base in the middle of the Kuwait desert, and what a great impact he made on all of us.

The next person to come was Vince Vaughn. This was right when the movie Old School came out. You better believe I didn't want to miss this meet and greet. The meeting spot was in a tent that was right next to our company headquarters tent. For a brief moment in time, Vince and I were tent neighbors...

Again, there was a long line, and we got to shake his hand. This time they had a camera person. High tech! Then later they shared the photos with us somehow. Since Old School just came out, people were doing the 'ear muffs' and covering their ears in the photo with him. He was doing it, too. When I walked up to him, the first thing I realized was how tall he was. I was like up to his elbow. I just smiled in the photo. Then I went back to my tent (neighbors remember?)

and I printed out a photo of him that I quickly found on a search engine. That sounds a little creepy but hey, my life there was seriously mundane, and this was awesome. I went back to the meeting tent, and they were packing up. The woman who was in charge of the tour didn't look happy and said they were done. I asked if he could please sign my sad little paper I just printed out from the printer in our OPS tent, and he said yes!

WHILE WE WERE GETTING ready to return home, I missed a great opportunity to meet James Gandolfini. He came to visit Doha, the main base in Kuwait, but he wasn't coming to our base. People from our battalion went to Doha for one reason or another every day. Our leadership got their hair done there for God's sake. People went there to wash their laundry and also shop at the PX. Since everyone was back from Iraq at this point, including my company commander, I was now the company executive officer, so therefore I had to ask permission to go places. So, I asked my company commander if I could go... and he said no. His reason: not everyone could go. Well not everyone wanted to go. Also, I had some connections with James Gandolfini: he practically comes from my hometown in north New Jersey, he went to high school with my second cousin, and we both went to Rutgers University! I was really disappointed and depressed when the company commander said that I couldn't go. I can remember how that deep emotion of disappointment felt through my body. Honestly, no one would have cared if I went; they would have been happy for me. This was just another tick on the list of the times the company commander was able to treat me like shit. Looking back, I shouldn't have asked and just went, then faced the consequences later. #WorthIt.

MAIL MATTERS

Mail downrange was a huge deal! We looked forward all day for the person doing the mail run to come back, and we could only hope that we had a letter or (even better) a package.

I wrote to a ton of different companies and asked them to please send some care items to raise the morale of the soldiers in my platoon. Sending mail was free for us, so I didn't have to buy any stamps, I just bought a box of envelopes and some writing paper. During one of my write-a-thons, I somehow realized that I was spelling a word wrong in all of my letters. I said it was for my platoon's moral. For their MORAL! When I first realized my mistake, I just froze with pen in hand, thinking how many people did I send out letters to asking to help my platoon's moral? Wow, they must have taken pity on me for not being the brightest bulb in the box. This LT isn't the sharpest tool in the shed. Maybe they sent more packages because of this pity? Or maybe that's why I got so many pocket dictionaries in those packages?

When I received the care packages, I would put them on a table in the OPS tent for everyone to take items. A lot of companies sent some really nice care and morale items. Frito-Lay sent a big pack of coupons for free bags of chips and I passed them out to everyone, which was greatly received. Some companies wrote back and said that they wanted to support us, but they couldn't, etc. I appreciated a response, though.

Looking back this seems crazy that I wrote to different organizations and companies, but I have to remember that we had all day to keep ourselves busy. And back then, mail really mattered! I would get addresses off of random food boxes that we had already and the soldiers liked. The internet didn't have so much information available yet.

However, online there were some sites that you could sign your unit up to receive care packages from organizations / groups back in America. Most of the groups were run by local churches, and families would sponsor us and send a package. I signed up my platoon for as many as I could. They received awesome t-shirts, snacks, games, and so much more. Once someone sent a massive box of rice Krispy Treats, and it was the best day ever. We didn't have them in the PX and everyone really loved having them.

I always made a point to send a thank you back to them! They had OIF and Kuwait post cards in the PX, so I would send them one of those. "Greetings from Kuwait!" "Howdy from OIF War."

The USO in Germany was really great, and I was friendly with the director already. We once got a conex sent to us filled with goods to hand out. Unfortunately, some Kuwaitis, or people in Kuwait, got to it first, broke the lock, and ravaged it. I never knew what else was in that conex before the ravaging.

From that delivery we got a ton of battery-operated spinning toothbrushes. Probably a good idea with all the damn snacks we were eating. Also, we got these cool looking electric razors for our females. Yeah baby.... downrange *and* still having silky smooth legs. Niiice. However, back in my room I learned that they weren't so nice, because the cool looking razor was actually an epilator, which spins and rips the hair out of your skin. I didn't know what the word epilator meant. I guess my spelling and vocabulary weren't the best. Because: Holy shit, OUCH!! Who uses that?! It was like a torture device. It should have been a clue that the conex thieves didn't even want that thing for free.

Once we got a package, with no return address on it, that was addressed to, "Alpha Company Commander, Signal Unit."

"Humm.. sounds like me, right? Let's open that bad boy up." (Dumb move #1)

Inside was a one gigabyte USB stick. Only a USB stick and nothing else. How strange and random, right? How many red flags would that be if I opened it now? Back then... not so much.

AND it was 1GB. This was HUGE in 2004. Nobody in the tent saw a USB stick with that much memory. I knew that I didn't order it from anywhere. So, talking to other people in the company, we thought maybe it was a gift. Riiight a gift...like the free potato chips. Sounds legit. So, we were keeping it. Let's try it out and put some files on it. A lot of files one it... one gigabyte!!!! Think of all the Excel sheets we can fit on that baby. Maybe even a few photos. I felt freakin' powerful. One gigabyte powerful.

So, word got around to a warrant officer, a Signal Corps warrant officer, that we had this 1GB USB. (These are the technical smart guys!) He came to my tent and offered me to swap his 1GB SD card for my new 1GB USB stick.

Deal!

I wanted to use it for my camera. I was also able to use it on my dinosaur, Army-issued laptop, to save at least four PowerPoint presentations.

Years later, we would actually be very smart about the dangers of plugging in unknown USB sticks into computers. In fact, now we can't even bring USB sticks into the same room we are working in. What we did back then was the textbook version of what not to do. China, Russia, Iraq... whoever, could have sent that with malware on it and when we stuck it in our computer, it could put a virus on our entire network! Shit, maybe it did!?

Semper Fi

Another hot Kuwaiti sandstorm. We were sitting in our OPS tent, staying away from the sandy madness that was outside. I was used to the tent shaking from the wind and hearing the canvas slap against our plywood walls, but I heard something different outside. I walked around the side of our OPS tent and there was a young Marine rummaging through a pile of wood we had stacked outside our tent. He was holding what he found: our old wooden door which blew off in a stand storm and we recently replaced with an upgrade.

When he saw me he immediately stood up, apologized, and stood at attention. I told him there was no problem, and I was wondering what he was looking for. He told me he would like our door for his colonel's tent. I found out he was the driver for the Marine battalion commander.

Units came in and out of Kuwait as a staging area before they drove on to Iraq. They wouldn't be there long enough to set up any supply accounts. We were getting ready to leave Kuwait soon and we couldn't take any of our supplies with us, so I told him to meet me and my supply sergeant in a few hours out by our conexes.

My supply sergeant and I took the Gator and drove out to meet the Marines in the desert to where we kept our big metal containers. They came in a couple of large Humvees. We unlocked the conex and told them they could have anything they wanted... in fact they could have it all. I admired the Marines and their professionalism, and I knew they had a hard road ahead of them in Iraq. I was glad to help them out in any way I could.

They were so appreciative; they said they did not want to just take everything from us and that they would have to make a trade somehow to repay us. Earlier I noticed some of their amphibious tanks, and I thought how cool it would be to go inside one. I asked them if this would be possible, and they agreed to give us a class on their badass tanks to my entire company and also throw in a hand-to-hand combat class. Hoo-rah!

So, a couple of days later my platoon sergeant got the company together for a formation and the entire Marine company came as well! We got a tour and lesson of the Marines' Expeditionary Fighting Vehicle (EFV). I sat in the driver's seat, and I was able to try the night vision camera they use for targeting. It was like a lens from an old submarine video where you pull down and can see what lurks out there. However, this had amazing zooming power! I was able to zoom across the desert miles away, and I saw a person outside a latrine, hand washing their clothes!

After that we had the hand-to-hand combat classes. Here we learned how to beat some ass and also how to hold our ground if a knife fight ever arose.

The Marine teachers were in the center of our circle; they would show us a move and then we would mimic them. The only difference between them and us was that the instructors were seriously beating each other... there was very little simulation.

Also, I found the knife fight lessons very handy. I learned, if someone is coming at you with a knife, and you are lucky enough to have one as well, the first thing to do is slam your arms simultaneously into the forearm of the arm that is trying to stab you. Then step in close to them and wrap your arm that is closest to knife, around that arm. This constricts any thrusting/ stabbing movement toward you. Then when you have their full torso exposed, you stab them. The Marine noted to first slice their biceps (so they can't attack/stab you) then go for their vital organs and drag the knife across the body like a dishonored samurai.

It was a great experience for our company because we would never have the opportunity to get that kind of training anywhere else.

When I said goodbye to my new Marine friend who I met trying to take my old door, he said that he was going to get a coin from his battalion commander and mail it to me. I thought this was a nice thing to say, but really, after fighting a war and then going back to California, he wouldn't remember that he even said this.

Eight months later in Germany, I received a package with a return address label from the Marines. I was shocked and so impressed that he remembered to send me the coin.

From this encounter, I learned the value of working with others and exchanging knowledge. Also... your word is your bond. I gained, and always will have, so much respect for Marines.

Leaving Kuwait

Our year was up, and it was time for everyone to go home. What a year it had been. When you really think about it, a year is a LONG time. It was hard to wrap my brain around being somewhere like that for a year. But we did it. We did it as a team. The big reason that we were deployed there might not have been clear for everyone, but we knew our unit's mission, and even more we did it because we were all there to support each other. We were a strong team and we were dedicated to support our brothers and sisters to our left and right in formation. We were all going home together, and that was what mattered the most. We were Renegades, and I will always be proud to have been "their LT" during this crazy year of our lives.

OUR BATTALION still stayed in Udari until the very last minute to meet the brigade in Doha for our contracted commercial flight home. We had Thanksgiving in the desert before we left. I thought it was awesome! But then again, I didn't have a family waiting for me at home. I would most likely be eating at the dining facility in Germany for Thanksgiving anyway. The food was the best they had, and they tried to make everything fancy, like little shrimps in a bowl of ice. We never had anything like that. What blew my mind was having a full bird colonel hand little ol' me a plastic champagne glass of sparkling cider. I felt like royalty.

Knowing that we were going home was exciting! For a single lieutenant, I was just happy to have my own life back again, my own bathroom again, *insert heavens opening sound here* and be able to travel again. Other married people felt something that I wasn't aware of yet: seeing their spouses again and seeing their kids again! I never really fully understood that feeling until I had kids of my own. Looking back, I can now feel how they must have been suffering. I miss my kids now while I am at work for eight hours, I can't imagine leaving them for a year.

We were at Doha for a week or so while we waited for our flight. We had nothing to do but just stay out of trouble and not go insane from being bored. Someone on the base charged about $100 to hack an Xbox 360 and load tons of games on it. So, we had a lot of new addicted gamers. Halo was very popular at the time. My soldiers would run Cat5 cable in the ceiling and connect their Xboxes so they could play together. You had to be creative when you didn't have any internet in the rooms.

I don't even know what we accomplished that week. A whole lot of nothing, I guess. We still stuck to our platoons and didn't really mix so much with the other companies. Maybe because we all felt like everything was coming to an end, and we wanted to emotionally wrap it all up together.

I was the company XO, but I had very close relationships with my original platoon. I still spent most my time with them, but I also got to know other soldiers from the headquarters and the other platoon. Like I said, we were borderline ready to snap from being bored, so one of the soldiers from First Platoon, Staff Sergeant P, and I stumbled upon Bravo Company's guidon left in the transient bunkbed rooms. This is totally taboo. As the guidon bearer, you always make sure you bring the guidon with you everywhere you go because any other company will take it and make you look very foolish. We weren't mean like other units, but we were going to do something. We took it to a contracted vendor who was by the PX making photographs of people dressed up like a Kuwaiti in front of various backgrounds. Staff Sergeant P and I decided it would be funny to take a photo together and have the guidon in the photo, print it out, and tell them we had their guidon held hostage. Looking back, being held hostage was a sensitive subject, not funny, and a little "too soon" if you know what I mean.

The woman dressed us up and told me that I would have to sit on the floor because I was a woman and I couldn't be higher or equal to my "husband" Staff Sergeant P. Okay whatever, this photo is absolutely ridiculous, and I can't believe people were paying for this for real. I naturally paid for the photo, and it was very expensive for what we got: dressing up for one printed photo, and me being subservient to a man for about thirty seconds. I didn't want my face in the photo so I pulled up the face opening on my body wrap so you couldn't see my mouth. I was laughing so hard when they took the photo!

We left the photo on the same bunk we found the guidon. So, the poor private or specialist who was in charge of the Bravo Company guidon would eventually retrace his steps throughout the day and end up at the bunk to find this photo. I thought it was pretty funny. I still do.

We then gave the guidon back. We were Signal Corps soldiers, and that was the extent of our savagery.

GET ON THE BIRD

The plane arrived, and we were all SO ready get the hell out of Kuwait. No more hurry up and wait, no more sand, no more humid hotness, no more terrible food, no more hours of hacked Xbox, no more shared bathroom bullshit. Home.

The whole brigade was getting our own contracted commercial plane. We were shuttled from station to station, and room to room. We had more hurry up and wait, more head counts, more dragging duffle bags. Everyone was up for a long time, and we were pretty exhausted. The staff loading the plane would call out the last four numbers of our social security numbers, and we had to YELL our names in this massive room. After a few names, people would start chatting and no one couldn't hear the guy saying the numbers or hear the soldier who was yelling their name. I mean, there were hundreds of people in this room. So, they would say last four, "2424!" and the soldier would yell, "Smith!" But the soldier who would go down in brigade history for at least the next few weeks was Donavan. He was near me, so I was able to hear him. The man yelled, "4242!" and Specialist Donavan said, "Donavan." He did. Then the super stressed and given-too-much-power civilian employee yelled with a lot of anger... "4242!!" And just Donavan's bad luck, somehow the entire hanger of the brigade went silent at the split second when Donavan replied, "Donavan, mother fucker!!" Picture hundreds of soldiers heads snapping quick to look at him in disbelief: WHAT did you just say?! Everyone laughed their asses off, but the brigade commander did not find that funny, and she looked at her command sergeant major and said, "Find out who that soldier was." This implied that the command sergeant major was to straighten his ass out later.

For the next few weeks, we all couldn't stop saying "Donavan, mother fucker!" It was like we all had Tourette's syndrome, and we physically couldn't stop our mouths from saying it in all various situations throughout the day.

Even on the plane going home, I dared one of my soldiers $20 if he would pick up the flight attendant phone and say, "Donavan, mother fucker." I probably should not have done that, but luckily he had some sense and didn't do it. Like I said, we were Signal nerds and not so savage.

Arriving "Home"

We landed in Ramstein Airbase because the Rhein Main Airbase in Frankfurt from which we deployed had closed. I didn't know that little tidbit of information when we landed, and surely didn't care at all. All we cared about was being home and out of the Middle East. We were shuffled from the plane to a bus like we were used to, so we didn't even see where we were, and I say again, we didn't care. I most likely fell asleep on the bus for the forty-five-minute drive to Mannheim. The families did not know we were coming home because that would have been a violation of operational security (OPSEC). Someone would have called them when we landed in Ramstein, and they would have a few hours to go to the base gymnasium for our arrival. They do this because if a terrorist knows when a large group of soldiers are moving together, we are one big target for an attack. So, everyone should take it very seriously.

Our bus pulled up, and we all marched right off the bus into the gymnasium. The families were all in the bleachers screaming and clapping! Everyone had signs for their soldier. The leadership called everyone to attention, then dismissed us. The family members screamed and ran into the formation to hug their soldiers. It was like a hallmark movie, with everyone is happy and crying. Well not everyone. I was there just happy to be home and shocked really to see "civilians" again, and they were all in "civilian" clothes. It had been a long time since seeing this and seeing women all dressed up looking pretty/sexy, while I felt like a tired, dirty, piece of trash.

The families cleared out of the gymnasium, the single soldiers were put onto another bus that took them to the barracks, which wasn't so far away. So, a little flaw in the welcome-home plans: where were the single officers supposed to go? I dragged my duffle bags outside thinking, wtf do I get a taxi or something? I want to go home, too. We didn't have our cars there obviously. They were all at our houses or in storage that was provided. The garrison commander was very nice, and she asked if we needed rides home. I said no thanks because it was awkward for me to say yes and drive with a full bird colonel for the next twenty minutes. Then, like a knight in shining armor, my lieutenant buddy, Siv, came by the welcome home ceremony and asked me if I needed a ride. He lived right in my town, so it was totally cool. I was so happy to see him after a year and was grateful that he took me home.

I got to my apartment to find that my car didn't start anymore, my electricity was turned off, my landlord saved my frozen food, and there were a lot of dead flies at the bottom of all the windows. However, I couldn't even deal with it then. Not so surprising, the first thing I did when I got home was go into my bathroom, (with a spin around to be extra like always), take a shower and then a much-needed nap.

CHAPTER 5

CPT

BACK FROM THE WAR

S hortly after coming back from Iraq, I was promoted to captain. Everyone in my company arrived home, and the unit went on block leave again, understandably. The equipment would take some weeks to return to us, and until then there was really nothing for us to do. I never joined the unit again after we arrived back. This was a very strange feeling, and it was hard for me because we went from being so extremely close for so long, and I was quickly not a part of them anymore. I was the executive officer, and this was a sought-after position, so it was my time to move on and make room for another lieutenant. Also, this was not a position for a captain anyway since you can't be the same rank as the company commander.

My battalion commander asked me what I would like to do next. I could have stayed at the battalion level and worked in a staff section, or I could have asked to go to the brigade (one level up) and work there. I really don't know why, but I wanted to shoot for the stars; I requested to go to the command level and work at 5th Signal

Command. I wanted to be different, and I wanted to challenge myself. It was my career, and these were my dreams. Like Bilbo Baggins said, "After all, why not?"

Little did I know that this decision would change my ENTIRE life and future career, and it is the reason why I have my current job. I got a position as a battle captain in the 5th Signal Command Center at Funari Barracks in Mannheim, Germany. Working there was a whole new world from what I was used to, on line with the soldiers. All the cursing had to stop for one thing. It was mostly senior NCOs and field grade officers. The newest aspect to get used to was working with the Department of Defense civilians. Working with them was so different; they might have as well been aliens from another planet.

I worked on shift in an operational command center. We briefed the G3, who was a colonel, and also the commanding general, who was a one-star general. We collected information from all of the subordinate units, created PowerPoint information decks, briefed them, and created situational reports to send out each day via email. We would also be prepared to brief the current status to any VIPS who would be visiting the command. We worked in a small room with lots of monitors and no windows. Being in there for twelve hours sometimes was not the best, but we had a nice schedule to make sure we had time off.

These were busy times because of two deployments going on at the same time. One unit was in Afghanistan, and one was preparing to go to Iraq. There were a lot of moving pieces, and there was no sign of it ever stopping soon. At first, I hated working there. I missed my soldiers, the team, and comradery. I was now in an office and doing death by PowerPoint to senior military leaders who didn't care that we just got back from Iraq. There was no mentorship, and my main boss was a civilian. He was great and all, but there was no mentorship for a young captain.

My shift work and lack of unit PT must have gotten to me and the scale because I broke tape. This meant that I failed the tape test.

Honestly, it was not by that much. I am so short, and the standard then was so ridiculous to me. I had to be counseled about this by a major who I worked with. I went in his office, and he said to me, "What hell is going on? It's calories in, calories out... it is not rocket science."

He made me feel so bad about myself and I was mortified standing there. What do you even say to that?

After that I starved myself, I went to the sauna in the gym with a plastic suit to sweat everything out, I worked out my forearm muscle because they taped this and it needed to be larger, and I weighed myself at least twice a day. Of course, I then lost the weight and passed... because... it is not rocket science like the a-hole major said.

Leadership changed, and over time I loved working there. I loved the DA civilians. They were like extended family. We did so many great hail and farewell events, holiday potlucks, BBQs, and after-work fests together. The 5th Signal Command Dragon Warriors were my new tribe and family. I still am connected with most of them to this day.

It was time for me to make a decision to either stay in or get out of the Army. I paid back Uncle Sam my four years, so anything after this was on me. I chose to not stay in the Army, and I was lucky enough to roll right into a DA civilian job there at 5th Signal Command. They knew me and my work, and they trusted me. I stopped working in the miliary uniform one day, went on leave in the States for a week, then came back to the same chair in civilian clothes, doing the same job on the same computer. I was twenty-six years old and already a GS-12. I was extremely lucky that my path somehow had gotten me to this job. If I didn't take the risk and leave my comfort zone, I would not have had this opportunity. Now I was one of the aliens from another planet, freaking out the new first lieutenants who started working there.

Chapter 6

Life After the Army

As Times Goes On

The deployment was over, and even my life in the Army was over, but it somehow never leaves you. When I talk about "war stories" with other veterans, I feel like I am talking about a different lifetime ago. Memories and details seem to get fuzzy as time passes by.

However, somewhere deep in my subconscious there are memories and echoes of the fear and feelings that I went through during my time in Baghdad. I came to this realization having my second child in 2017. I was not thinking about anything close to the Army or Iraq on the day I was having my scheduled c-section. I was focused and excited to meet our new daughter. I had to have the spinal numbing so that I did not feel the lower half of my body. I read that this drug can cause some gnarly dreams and/or nightmares. Once I had the baby, I was back in my room waiting to be able to walk again and I fell asleep for a few moments in bed. Out of nowhere I had the most horrible nightmare that shook me to my core.

In the dream, I was back in Iraq in the Shady Shop in the back of the base, and the Lebanese guy was behind the counter. Then with an evil smile on his face, he lifts up this big Ziplock bag and in it is the bottom half of the leg of the soldier that I saw get pulled off the flat bed! I woke up with a huge gasp and I was shaking. I was shocked and scared. Where the hell had such a horrible and gruesome thought came from? I was disgusted that I was even able to come up with something like that in my head. And... WHY?! I felt so horrible I was scared to fall back asleep. Also, who was I going to tell, who was going to understand me?

That is why it is important to keep in touch with fellow veterans. Buddy checks. Check on each other. You never know what they might be going through.

While you know that your same group of people will never be physically together again, you have already formed a bond that is there for life. You formed a band of brothers. You have their back, and you know that they have your back. As they say in the Army, "I got your six."

I use social media to keep in touch with a lot of the people I have served with. I enjoy seeing how the once young soldiers in my platoon are all grown up with their own families now. I am probably still "old" to them, and they still call me ma'am. In my head they are still young and baby-faced, but I see how time has aged them. We spent special time together in our youth that will forever bond us all. I see that the soldiers from our Second to None Platoon visit each other still. I love this! Twenty years later and the connection and desire to keep in touch is still there. There are a few of us who stayed in Germany, but most of the platoon is spread across the globe.

MY TRIBE

As time goes on, I have found my new tribe. As they say, your vibe attracts your tribe. Also, it is important to only surround yourself with people who charge your battery instead of those who drain it. I have found like-minded individuals who inspire me and, as I get older, I spend my time more wisely, choosing who I associate myself with. I am tired of saying yes to people and plans out of the fear of being impolite. Life is too short for this, and I have to really analyze how people make me feel after being with them.

I have found other leaders who build me up, and I support them - others who have the same drive to serve and to give back to the community. My tribe consists of friends who are also mentors, who can help guide me and give me feedback. A tribe lifts you up without any feelings of competition or jealousy. The tribe wants you to succeed.

There is no easy / textbook way to teach this because it is a feeling; you just know. You know what you are with your peeps. You know when you are cut from the same type of wood, and on the same wavelength. It is the feeling of positivity and being charged when you leave these people. Find your tribe of people who have the same goals and values as you do and make an effort to continue to work together and spend time together.

In the words of motivational speaker Jim Rohn, "You are the average of five people you spend the most time with."

PRIVATE ORGANIZATIONS AND STILL SERVING

Veterans, and like-minded people, have a strong desire to serve. It is important for veterans to find a place where they can still serve like they did in the military. After many years of being back from the deployment, and out of the Army completely, I still feel the desire to give back to the Army, to support America's soldiers and military families, and to honor the fallen. I have found the type of people I love being around: like-minded professionals who also want to volunteer their time to support the military. #StillServing #SoldierforLife

The Association of the United States Army (AUSA) is the Army's professional organization. It is a the "Voice of the Army – Support for the Soldier." Being involved has been such an incredible experience for me, and I have met the most wonderful people in this organization. So many doors and opportunities opened for me because I surrounded myself with these inspiring leaders. I have had opportunities to meet the Army's senior leaders and learn so much from them. I have learned and I have grown volunteering with my local chapter. I love the mission and supporting America's Army.

I took a photo with the leadership from my chapter and after looking at it a few days later, it dawned on me that it was so similar to the photo I took with my platoon on my last night in Baghdad. I guess some things don't change so much!

VETERANS OF FOREIGN Wars (VFW) of the United States is a nonprofit veteran's service organization comprised of eligible veterans and military service members. Their mission is to foster camaraderie among United States veterans of overseas conflicts; to serve our veterans, the military, and our communities; and to advocate on behalf of all veterans. I am a member of my local VFW Post and I believe it is important for Veterans to connect and still form a comradeship with others who have served overseas. I am proud of what VFW stands for, supporting veterans rights, exemplifying true patriotism, and comradeship. I also cherish the relationships I have made with the Vietnam veterans in my area and have learned so much from them about what they went through while serving overseas in Vietnam.

My grandfather was in the VFW right after WWII, and he was the one of the founding commanders in my hometown. He was really proud when I joined his VFW Post after I came back from Iraq. One year I went on a float with him in my hometown during a Memorial Day Parade and he was very happy to have me there. He showed me off to all of his battle buddies and told them that "his pretty granddaughter was an officer!"

He also joked how he could still fit in his Eisenhower jacket after all these years, and how it would still button up! Not many men could still wear the jacket anymore. When he passed away, my uncle was kind enough to give me his jacket, and it is now on display right in front of the Commanding General of US Army Europe and Africa's Office in the headquarters in Wiesbaden, Germany. Visitors and leaders from all over the world look at his photos and jacket in the display cabinet, and that makes me feel so proud.

FOR MY GRANDFATHER and all of the soldiers that he served with in WWII, I have made it a lifelong dedication to honor their duty and sacrifice in Europe. My friends, team, comrades, and I visit the historic locations and cemeteries. We honor them with wreaths and military commemoration events. Our AUSA chapter has a partnership with the incredible 101st Airborne Museum in Bastogne, Belgium, which consists of the best museum team members who are passionate about keeping history alive in honor of the US soldiers who fought there in 1944. We visit the foxholes of Easy Company from 'Band of Brothers' and place US flags to show respect. We visit Bastogne, Belgium, and participate in the remembrance and reenactment events during the NUTS weekend. We support all the D-Day events in Normandy, France. They shall not be forgotten.

We exemplify the motto of the American Battle Monuments Commission (ABMC): Time will not dim the glory of their deeds.

.

BIG LESSONS LEARNED

Here are some big lessoned that I learned throughout my whole leadership journey.

It's okay to not be perfect.

There is something powerful in being vulnerable and admitting you made a mistake. Always learn from your mistakes. After every single bump in the road, you must step aside and learn the lesson from the moment. Analyze and learn from the mistake. Be rid of any anger about the situation and be glad that it happened so you are able to be better next time. It will make you a stronger and more resilient person. You will improve yourself and those who are watching you.

Anyway, people like you better when you aren't perfect.

Be Reliable

The biggest hurdle is just showing up! If you said you are going to be somewhere, or do something, be the kind of person that everyone knows will show up or complete the task. If you can't make it or do it because of unforeseen circumstances, then just tell the person. Don't be a flake. People will stop relying on you. No one has time for flakes. Flakes annoy the hell out of reliable people and planners.

This kind of behavior will close more doors for you and not expand your professional network. I can't start to list all the opportunities that I have had and the people I have met because I have just shown up or followed through. Showing up might open another door to an opportunity and then another opportunity like a domino effect.

Find your Community

When soldiers leave the Army, they miss the comradery and the closeness with others who share the same values. This leads to feeling lonely and different, and sometimes leads to suicide. If you are a veteran, or you are a liked-minded civilian, find your tribe.

Once you found your type of tribe, then find your community. Find a community that you feel a part of where you can contribute.

Improve the community in any way possible. Either be the

person who brings people together or be the person who supports the community leaders.

Show up and volunteer if you are able to. Think about how the entire world would change for the better if everyone volunteered just a few hours each year. Think about a the time you volunteered in your community.

I learned this Chinese proverb when I was in a teen leadership group in high school and it has always resonated with me:

"If there is light in the soul,
There will be beauty in the person.
If there is beauty in the person,
There will be harmony in the house.
If there is harmony in the house,
There will be order in the nation.
If there is order in the nation,
There will be peace in the world."

VISIT *other Communities*

I feel like communities are gears, connected together with each other moving and spinning across the world. We are all somehow connected but really spinning in our own gear of what is affecting us, what is around us. I love to travel and see other communities/places to broaden my thinking and have a change from the norm.

CHALLENGE *Yourself*

This is where we grow.

Do things that scare you. I don't mean go bungie jumping or touch spiders. That's terrifying and please don't do those things. I mean to set your professional and life goals and then put yourself in situations where you are uncomfortable. The situations that make your heart race and push you have out of your comfort zone are where you grow! If you don't push yourself out there in this zone, you will stay exactly where you are, and as a leader you should always strive to continue to flourish.

What would you want to accomplish if you had no constraints?

Step 1. List all of the things that are stopping you from doing that goal. Take a good hard look at them.

Step 2. Really look at what is stopping you. Are they that bad? How can you work on getting past or solving each one?

Step 3. Do the damn thing.

"Twenty years from now you will be more disappointed by the things you didn't go than by the ones you did do. So, throw off the bowlines! Sail away from the safe harbor. Catch the trade winds in your sails. Explore. Dream. Discover." - Mark Twain

FLOWERS FROM BAGHDAD

It was one of the bloodiest and worst months of the war, so far, with the most casualties. Everything felt hard and the mood was heavy. I left my OPS space and headed to the brigade headquarters that we were attached to. To get there I had to walk to the large field where the company commander and first sergeant had their small building, then I turned right and walked along a long dusty road that followed the perimeter of the base. The sun was shining bright that day, and I needed my sunglasses and the brim of my hat to shield my eyes. I walked along by myself, and I was tired - emotionally and physically tired. I thought, "Isn't it strange how is the climate so vacation-like, but we are at war."

Then from a distance I heard a machine gun blasting in bursts, and it startled me. Then I heard more explosions coming from the same area. I already couldn't shake the feeling that everything with the attacks was getting so bad, and this was not helping. I continued to walk to the brigade headquarters, with my head down to make a shadow on my face. I was feeling low.

Then I looked over to the side of the dusty road and there were beautiful purple flowers growing! There was nothing else flourishing at all around, just these luscious deep violet flowers. I hadn't seen any flowers in a while, but I had been having this strange connection with purple lately when I fell asleep. Honestly, my mood was suddenly lifted.

Then a strong epiphany came over me seeing these flowers. My mind was getting dark and depressed, there were sounds of war in the background but seeing these flowers amidst all of that showed me that there is ALWAYS something positive to find in any bad situation. In all of this death and war, I still found beauty.

It taught me to always remember that in all situations in life you can always find the positive in the situation and focus on that, to turn your thinking into a positive attitude. I realized that I have the power to flip the script and not give in to the negativity and hopelessness. Even as a leader this is something very important to focus on because others are watching you and you are their role model. As a platoon leader, how do you think my soldiers' morale would have been if they saw their lieutenant down-in-the-dumps with a negative attitude? That kind of negativity spreads like a wildfire.

In all of my Army experiences and my leadership lessons, as a young lieutenant, I realized that I always did find the positive in the situations, and my soldiers saw that. I really hope that it somehow helped us get through it all.

Consider this lesson from my long journey to and from a horrible war overseas: Everything you have been through, and anything you experienced, no matter how dark the situation feels at the moment, you can stop and take control to change your thinking into a positive attitude. Find YOUR flowers from Baghdad.

ACKNOWLEDGMENTS

In memory of my mother, Ginny Fusco, who always believed in me and my writing. My number one fan!

I want to express my deep gratitude to David Fulton for writing an incredible foreword for Flowers From Baghdad! I am truly honored for his mentorship and inspiration over the years. His support means the world to me and his foreword was the icing on the cake of this book!

My heartfelt thanks go to my incredible family and friends who supported me along the way while writing the book!

Thank you to my battle buddy bestie, Stefan Deisenroth, for the constant support and always being a perfect sounding board!

Merci beaucoup to Johnny Bona and the entire team at the 101st Airborne Museum in Bastogne, Belgium, for their unwavering support. Commander Bona, thank you for the praise on the book cover! Baghdad to Bastogne!

Much gratitude to Erick Ocasio, CEO and founder of Leadership is Tricky for the wonderful praise on the book cover! I value the 'Leadership is Tricky' podcast, his friendship, and his mentorship very much!

Appreciate the guidance provided by ALL my friends and mentors who helped me create the final version of Flowers From Baghdad.

I wrote most of this book in a relaxed state of mind, in the incredible Alps in Bavaria, or the Hessen Countryside, while watching my daughters Maeve & Emily play. It has been extremely therapeutic, and I appreciate Tactical 16 Publishing for giving me, and other Veterans, this opportunity. Thank you to the entire team for the guidance, edits, and the perfect book cover!

EXTRAS

Pogs from AAFES. They used these instead of giving change because it was too heavy to transport. I can technically still take this in and cash them at the PX.

Money bought from the Shady Shop in Baghdad.

ABOUT THE AUTHOR

Hailing from New Jersey and now living in Germany, Gemma McGowan is a former US Army captain, a veteran, and currently a Department of Defense Army civilian. She holds a M.S. in Organizational Leadership from Southern New Hampshire University and a B.A. in American Studies from Rutgers, the State University of New Jersey. She also attended a post-graduate seminar class at the Harvard University, Kennedy School and a Leader Development Program at the Center for Creative Leadership in Brussels, Belgium. She received a Bronze Star Medal for her leadership and actions in Baghdad, Iraq. She was the first female VFW Post Commander for VFW Post 27, Watch on the Rhine in Wiesbaden, Germany, and is currently the Association of the United States Army (AUSA) Chapter President of the GEN C.W. Abrams Chapter in Wiesbaden, Germany.

About the Publisher
TACTICAL 16

Tactical 16 Publishing is an unconventional publisher that understands the therapeutic value inherent in writing. We help veterans, first responders, and their families and friends to tell their stories using their words.

We are on a mission to capture the history of America's heroes: stories about sacrifices during chaos, humor amid tragedy, and victories learned from experiences not readily recreated — real stories from real people.

Tactical 16 has published books in leadership, business, fiction, and children's genres. We produce all types of works, from self-help to memoirs that preserve unique stories not yet told.

You don't have to be a polished author to join our ranks. If you can write with passion and be unapologetic, we want to talk. Go to Tactical16.com to contact us and to learn more.

All of Tactical 16's books are available on our online bookstore, T16Books.com. Visit it today to see more books from our selection of authors and to find a new adventure to read!

www.ingramcontent.com/pod-product-compliance
Lightning Source LLC
Chambersburg PA
CBHW060926120626
46557CB00003B/892